LECTURE NOTES ON COMPOSITE MATERIALS

T0181347

SOLID MECHANICS AND ITS APPLICATIONS
Volume 154

Series Editor: **G.M.L. GLADWELL**
Department of Civil Engineering
University of Waterloo
Waterloo, Ontario, Canada N2L 3GI

Aims and Scope of the Series

The fundamental questions arising in mechanics are: *Why?, How?,* and *How much?*
The aim of this series is to provide lucid accounts written by authoritative researchers giving vision and insight in answering these questions on the subject of mechanics as it relates to solids.

The scope of the series covers the entire spectrum of solid mechanics. Thus it includes the foundation of mechanics; variational formulations; computational mechanics; statics, kinematics and dynamics of rigid and elastic bodies: vibrations of solids and structures; dynamical systems and chaos; the theories of elasticity, plasticity and viscoelasticity; composite materials; rods, beams, shells and membranes; structural control and stability; soils, rocks and geomechanics; fracture; tribology; experimental mechanics; biomechanics and machine design.

The median level of presentation is the first year graduate student. Some texts are monographs defining the current state of the field; others are accessible to final year undergraduates; but essentially the emphasis is on readability and clarity.

For other titles published in this series, go to
www.springer.com/series/6557

Lecture Notes on Composite Materials

Current Topics and Achievements

Edited by

RENÉ DE BORST

Eindhoven University of Technology
Eindhoven, The Netherlands

and

TOMASZ SADOWSKI

Lublin University of Technology
Lublin, Poland

 Springer

Editors

René de Borst
Department of Mechanical Engineering
Eindhoven University of Technology
Eindhoven
The Netherlands
R.d.Borst@tue.nl

Tomasz Sadowski
Department of Applied Mechanics
Faculty of Mechanical Engineering
Technical University Lublin
ul. Nadbystrzycka 36
20-618 Lublin
Poland
t.sadowski@pollub.pl
archimedes.pol.lublin.pl

ISBN 978-90-481-7981-7   e-ISBN 978-1-4020-8772-1

Cover illustration: WMXDesign GmbH

Printed on acid-free paper.

9 8 7 6 5 4 3 2 1

springer.com

Preface

Composite materials are heterogeneous by nature, and are intended to be, since only the combination of different constituent materials can give them the desired combination of low weight, stiffness and strength. At present, the knowledge has advanced to a level that materials can be tailored to exhibit certain, required properties. At the same time, the fact that these materials are composed of various, sometimes very different constituents, make their mechanical behaviour complex. This observation holds with respect to the deformation behaviour, but especially with respect to the failure behaviour, where complicated and unconventional failure modes have been observed.

It is a challenge to develop predictive methods that can capture this complex mechanical behaviour, either using analytical tools, or using numerical methods, the finite element method being the most widespread among the latter. In this respect, developments have gone fast over the past decade. Indeed, we have seen a paradigm shift in computational approaches to (composite) material behaviour. Where only a decade ago it was still customary to carry out analyses of deformation and failure at a macroscopic level of observation only – one may call this a phenomenological approach – nowadays this approach is being progressively replaced by multiscale methods. In such methods it is recognized a priori that the overall behaviour is highly dependent on local details and flaws. For instance, local imperfections in spacing and direction of fibres can be detrimental to the overall bearing capacity of a structure that is composed of such a fibre-reinforced composite material. By upscaling, homogenization or methods that in a single calculation take into account the behaviour at different scales, an attempt is made to design numerical methods that have a wider range of applicability – by less reliance on adhoc assumptions – and are better rooted in the true physical behaviour of the constituent materials. Yet, few monographs have been published that present an account of recent developments in the analytical/numerical modelling of composite materials.

This volume – which has grown out of a series of lectures that has been given at Lublin University of Technology within the framework of the European

Community Marie-Curie Transfer-of-Knowledge project *Modern Composite Materials Applied in Aerospace, Civil and Sanitary Engineering: Theoretical Modelling and Experimental Verification* (contract MTKD-CT-2004-014058) – aims to fill this gap. It starts by a comprehensive account of methods that can be used at macroscopic level, followed by a précis of recent developments in modelling the failure behaviour of composites at a mesoscopic scale. Going down further, the third chapter treats fundamental concepts in micromechanics of composite materials, including the essential concept of the Representative Volume Element and Eshelby's method. As recognized widely, failure is seldom a consequence of pure mechanical loadings. Often, thermal effects and long-term effects for instance due to hygric or chemical actions play an important role as well. For this reason the ensuing two chapters are devoted to thermal shocks and the numerical treatment of diffusion phenomena in addition to mechanical loadings when describing failure in heterogeneous materials. The volume is completed by a review of fracture mechanics tools for use in the analysis of failure in composite materials.

Eindhoven and Lublin,
René de Borst and Tomasz Sadowski

Contributing Authors

Holm Altenbach is a Full Professor of Engineering Mechanics and the Director of the Center of Engineering Sciences at the Martin-Luther-Universität Halle-Wittenberg (Germany). His research interests are focussed on the following topics: Structural Mechanics (beams, plates and shells), Lightweight Structures (laminates and sandwiches), Continuum Mechanics (basics and constitutive modelling) and Creep-damage Analysis. Since 2005 he has become one of the Editors-in-Chief of the Zeitschrift für angewandte Mathematik und Mechanik.

René de Borst is a Distinguished Professor at Eindhoven University of Technology and a member of the Royal Netherlands Academy of Arts and Sciences. His current research interests are in the development of novel numerical methods for the analysis of multiscale phenomena, multiphysics problems, and evolving discontinuities.

Eduard Marius Craciun is a Professor at the Faculty of Mathematics and Informatics at the "Ovidius" University of Constanta (Romania). Main research field represents the analytical and numerical methods in the study of crack propagation (incremental values of stresses producing crack propagation and crack propagation direction) in prestressed elastic composites and in prestressed and prepolarized piezoelectric materials.

Ryszard Pyrz is Professor and Chair in Materials Science and Engineering and a head of the Center for New Era of Materials Technology (NEMT) at the University of Aalborg (Denmark). His scientific interest and research activities comprise following areas: molecular modelling of nanomaterials and nanostructures, development of a direct connection between processing and microstructure of advanced materials, multiscale modelling methodologies, experimental *in situ* investigation of materials' microstructure with utilization of modern techniques such as Raman nano/microspectroscopy and scanning probe microscopy.

Tomasz Sadowski is the Head of the Department of Solid Mechanics at the Lublin University of Technology (Poland). Prof. Sadowski's research interests comprise the following areas: continuum damage mechanics of materials and structures, modelling of ceramic polycrystalline materials, modelling of composites with ceramic and polymer matrix, fracture mechanics of materials under mechanical loading and thermal shock.

Contents

ANALYSIS OF HOMOGENEOUS AND NON-HOMOGENEOUS PLATES

Holm Altenbach
Lehrstuhl für Technische Mechanik
Zentrum für Ingenieurwissenschaften
Martin-Luther-Universität Halle-Wittenberg
D-06099 Halle (Saale)
Germany
holm.altenbach@iw.uni-halle.de

Abstract   Plate theory is an old branch of solid mechanics – the first development of a general plate theory was made by Kirchhoff more than 150 years ago. After that many improvements were suggested; at the same time some research was focussed on the establishment of a consistent plate theory. Plate-like structural elements are widely used in classical application fields like mechanical and civil engineering, but also in some new fields (electronics, medicine among others). This paper gives a brief overview of the main theoretical directions in the theory of elastic plates. Additional information is available in the literature.

Keywords: structural analysis, plates, homogeneous and non-homogeneous cross-sections

1.   Classification of structural models

Plates are structural elements with applications in various branches. The reason for this is that plates combine high bearing capacities with low weight (excellent specific stiffness properties). Modern plate structures are made from different materials – it is common to use classical structural materials like steel or concrete, but also modern composite materials like laminates. Increasing safety requirements dictate necessity of improving the analysis of plates. Since all commercial *Finite Element* codes allowing their analysis have special plate elements, but the manuals do give not enough theoretical background, an overview of the modeling approaches in the plate theory will be given.

R. de Borst and T. Sadowski (eds.) *Lecture Notes on Composite Materials – Current Topics and Achievements*

1.1   Introductional remarks

The basic problems in engineering mechanics are the analysis of strength, vibration behavior and stability of structures with the help of structural models. Structural models can be classified, for example, by their

- Geometrical (spatial) dimensions

- Applied loads

- Kinematical and/or statical hypotheses approximating the behavior

A complex structure can be built up of many individual structural elements; the behavior of the whole structure includes the interaction of all parts.

Let us introduce three basic classes of structural elements. The *first* one is the class of three-dimensional structural elements which can be defined as follows:

> *A three-dimensional structural element has three spatial dimensions of the same order; there is no predominant dimension.*

Typical examples of geometrically simple, compact structural elements in the theory of elasticity are shown in Fig. 1.

The *second* is the class of two-dimensional structural elements which can be defined as follows:

> *If two spatial dimensions have the same order, and the third, which is related to the thickness, is much smaller, one has a two-dimensional structural element.*

Typical examples of two-dimensional structural elements in civil engineering/structural mechanics are shown in Fig. 2.

The *last* class is related to one-dimensional structural elements which can be defined as follows:

> *Two spatial dimensions, which can be related to the cross-section, have the same order. The third dimension, which is related to the length of the structural element, has a much larger order in comparison with the cross-section dimension orders.*

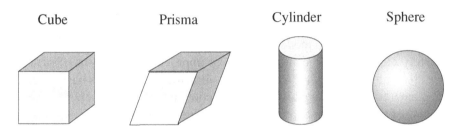

Figure 1. Examples of simple three-dimensional structural elements

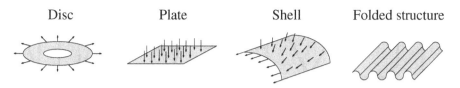

Figure 2.   Examples of simple two-dimensional structural elements

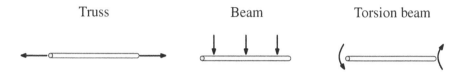

Figure 3.   Examples of simple one-dimensional structural elements

Typical examples in engineering mechanics are shown in Fig. 3.

It is possible to introduce other classes. For example, in shipbuilding, thin-walled structural elements are often used. These are thin-walled light-weight structures with a special profile, and they require an extension of the classical one-dimensional structural models:

> *If the spatial dimensions are of significantly different order and the thickness of the profile is small in comparison to the other cross-section dimensions, and the cross-section dimensions are much smaller in comparison to the length of the structure one can introduce quasi-onedimensional structural elements.*

Suitable theories for the analysis of quasi-onedimensional structural elements are:

- Thin-walled beam theory (Vlasov-Theory; Vlasov, 1958) and

- Semi-membrane theory or generalized beam theory (Altenbach et al., 1994)

Typical thin-walled cross-section profiles are shown in Fig. 4.

1.2   Two-dimensional structures – definition, applications, some basic references

Let us introduce the definition of a two-dimensional structure:

> *A two-dimensional load-bearing structural element is a model for analysis in Engineering/Structural Mechanics, having two geometrical dimensions which are of the same order and which are significantly larger in comparison with the third (thickness) direction.*

Closed cross-section profiles

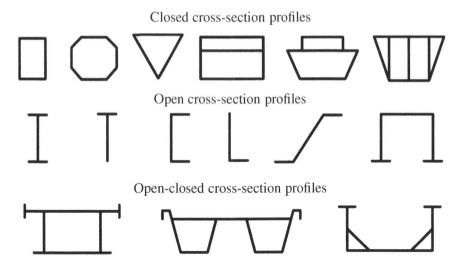

Open cross-section profiles

Open-closed cross-section profiles

Figure 4. Various profiles of thin-walled structures

This definition does not contain any restriction on the type of loading (in-plane, transverse, etc.). In addition, the thinness hypotheses (see, e.g., Başar and Krätzig, 1985) is not specified.

The mathematical consequence is obvious: instead of a three-dimensional problem, which is represented by a system of coupled partial differential equations with respect to three spatial coordinates, one can analyze a two-dimensional problem, which is described by a system of coupled partial differential equations with respect to two spatial coordinates. The two coordinates represent a surface; the behavior in the thickness direction is approximated mostly by use of engineering assumptions. The transition from a three-dimensional to a two-dimensional problem is non-trivial, but the solution effort decreases significantly and the possibility of solving problems analytically increases.

Two-dimensional structures have many applications in various branches: thin homogeneous plates, thin inhomogeneous plates (laminates, sandwiches), plates with structural anisotropy, moderately thick homogeneous plates, folded plates, membranes, biological membranes, etc. The main industrial branches for plate applications are aeronautics and aircraft industries, automotive industries, shipbuilding industries, vehicle systems, civil engineering, medicine,
. . . .

During the last few years many scientific papers, textbooks, monographs and proceedings about the state of the art and recent developments in plate theory have been published. Some of the most important publications are listed here without comment.

- Review articles: Naghdi, 1972; Grigolyuk and Kogan, 1972; Grigolyuk and Seleznev, 1973; Reissner, 1985; Noor and Burton, 1989a,b; Noor and Burton, 1990a,b; Reddy, 1990; Irschik, 1993; Burton and Noor, 1995; Noor et al., 1996

- Monographs and textbooks: Panc, 1975; Kączkowski, 1980; Girkmann, 1986; Timoshenko and Woinowsky Krieger, 1985; Ambarcumyan, 1987; Gould, 1988; Reddy, 1996; Altenbach et al., 1998; Woźniak, 2001; Zhilin, 2007

- Actual conferences like EUROMECH 444 (Kienzler et al., 2004), Shell Structures Theory & Applications (Pietraszkiewicz and Szymczak, 2005), IUTAM Symposium Relation of Shell, Plate, Beam and 3D Models (Jaiani and Podio-Guidugli, 2008)

1.3   Formulation principles, historical remarks

The plate equations can be deduced as follows (Altenbach, 2000b; Wunderlich, 1973):

- Starting from a 3D continuum and

- Starting from a 2D continuum

If one starts from the 3D continuum there are two possibilities:

- The use of hypotheses to approximate the three-dimensional equations and

- The use of mathematical approaches (series expansions, special functions, etc.) to develop a set of two-dimensional equations

All these methods have advantages and disadvantages and it is difficult to say in advance which is the best method for derivating the governing equations. In addition, it can often be shown that different methods lead to identical sets of equations.

Engineers prefer theories which are based on hypotheses. For example, there are many theories which are based on displacement approximations. Let us consider the plate geometry as shown on Fig. 5. The three displacements u_i in the classical three-dimensional continuum are now split into in-plane displacements u_α and the deflection w. The first theory of plates based on displacement assumptions, was presented by Kirchhoff, 1850. The modern form of the basic assumptions, which can be used for homogeneous and non-homogeneous plates, is

$$
\begin{aligned}
u_1(x_1, x_2, z) &\approx u_1^0(x_1, x_2) - z w_{,1}(x_1, x_2), \\
u_2(x_1, x_2, z) &\approx u_2^0(x_1, x_2) - z w_{,2}(x_1, x_2), \\
w(x_1, x_2, z) &\approx w(x_1, x_2)
\end{aligned}
\tag{1}
$$

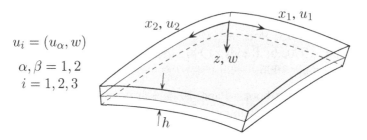

$u_i = (u_\alpha, w)$

$\alpha, \beta = 1, 2$

$i = 1, 2, 3$

Figure 5. Plate geometry and displacements

or using the Einstein's summation convention

$$u_\alpha(x_\beta, z) \approx u_\alpha^0(x_\beta) - zw_{,\alpha}(x_\beta), \qquad w(x_\beta, z) \approx w(x_\beta), \qquad (2)$$

where $u_\alpha(x_\beta, z)$ are the three-dimensional displacements, $u_\alpha^0(x_\beta)$ are the displacements of the reference surface (usually this surface is assumed to be the mid-plane) and $w(x_\beta, z)$ are the three-dimensional deflections which are approximately equal to the two-dimensional deflections $w(x_\beta)$. $(\ldots)_{,\alpha}$ denotes the derivative with respect to the in-plane coordinates x_α.

Approximately 100 years later this theory was improved (see, for example, Hencky, 1947; Mindlin, 1951)

$$u_\alpha(x_\beta, z) \approx u_\alpha^0(x_\beta) + z\varphi_\alpha(x_\beta), \qquad w(x_\beta, z) \approx w(x_\beta) \qquad (3)$$

In this equations $\varphi_\alpha(x_\beta)$ are the cross-section rotations. Comparing both approaches, one can see that the improvement was realized by introducing additional degrees of freedom. Calculating the strains as usual in the theory of elasticity, one gets

- Neither theory takes into account the thickness changes and

- The Kirchhoff theory leads to zero transverse shear, while the improved theory considers transverse shear in an approximate sense

The introduction of independent rotations is in some cases not enough, since it is assumed that any cross-section will be plane before and after deformation. For example, for plates made from rubber-like materials, the assumption of plane cross-section is not valid. A weaker assumption was proposed by Levinson (1980) and Reddy (1984) among others

$$u_\alpha(x_\beta, z) \approx u_\alpha^0(x_\beta) - [w_{,\alpha}(x_\beta) + \varphi_\alpha(x_\beta)]\frac{4z^3}{3h^2}, \qquad w(x_\beta, z) \approx w(x_\beta)$$
$$(4)$$

The latter representation and the Kirchhoff or Mindlin plate equations can be discussed from the point of view of introducing additional degrees of freedom.

On the other hand all equations of this type can be understood as some part of a power series. The first suggestion of this type was made by Lo et al., 1977a. Some kind of generalization of the power series approach was given in Meenen and Altenbach, 2001

$$u_\alpha(x_\beta, z) = \sum_{q=0}^{K_1} u_\alpha^q(x_\beta)\phi^q(z) + \sum_{q=0}^{K_2} w_{,\alpha}^q(x_\beta)\psi^q(z),$$

$$w(x_\beta, z) = \sum_{q=0}^{K_2} w(x_\beta)^q \chi^q(z)$$

(5)

The disadvantage of this approach is that, with an increasing number of terms in the series the physical interpretation of all terms is impossible.

In addition, the method of hypotheses for the stress and/or the strain (displacement) states was applied in Reissner (1944, 1945, 1947), Bollé (1974a,b) and Kromm (1953). It is easy to show that, for example, Mindlin's and Reissner's theories contain partly identical equations, but the coefficients differ slightly, and the interpretations are not the same.

Purely mathematical approaches are mostly based on power series, trigonometric functions, or special functions, etc. (see, e.g., Lo et al., 1977a,b; Kienzler, 1982; Preußer, 1984; Touratier, 1991). The mathematical approaches are very helpful if one wants to check the accuracy of the given approximation. A nice comparison of the different approximations in the series approach is given in Kienzler (2002).

The direct approach is based on the *a priori* introduction of an two-dimensional deformable surface. This approach was applied by Günther (1961), Green et al. (1965), Naghdi (1972), Rothert (1973), Zhilin (1976, 1982), Palmow and Altenbach (1982), Robin (2000), etc. This approach is still under discussion since the application is not trivial; but the direct theories are mathematically and physically so strong and as exact as three-dimensional continuum mechanics.

2. Classical plate theories

Below, we discuss some aspects of classical plate theories. Classical plate theories are theories based on hypotheses and which are mostly used in engineering practice.

2.1 Small deflections

In contrast to other classifications, here we first discussing two types of models: small and large deflection models. There are two basic theories used in practice: the Kirchhoff and the Mindlin theories. In the simplest case, both are

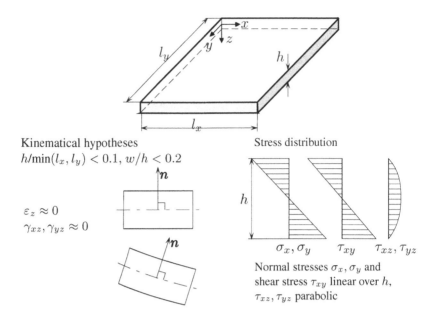

Kinematical hypotheses

$h/\min(l_x, l_y) < 0.1,\ w/h < 0.2$

$\varepsilon_z \approx 0$

$\gamma_{xz}, \gamma_{yz} \approx 0$

Stress distribution

Normal stresses σ_x, σ_y and
shear stress τ_{xy} linear over h,
τ_{xz}, τ_{yz} parabolic

Figure 6.   Kirchhoff plate – basic assumptions

restricted to small deflections (that means less than 0.2 of the plate thickness). The basic assumptions of Kirchhoff plate theory are shown in Fig. 6. They can be summarized as follows: no thickness changes, no transverse shear, linear distributions of the in-plane stresses, and parabolic distribution of the transverse shear stresses. The model can be applied to plates made of classical isotropic materials, and for small deflections; it is assumed that any cross-section it must be plane and orthogonal to the mid-plane before and after deformation.

Since the Kirchhoff model omits transverse shear, the Mindlin model is a suitable improvement. Now the assumption of plane cross-sections before and after deformation holds valid, but the cross-sections are no longer orthogonal after deformation. This assumption leads to additional degrees of freedom – two independent rotations. The basic assumptions of the Mindlin plate theory are shown in Fig. 7. They can be summarized as follows: no thickness changes, constant transverse shear, linear distributions of in-plane stresses, and constant distribution of the transverse shear stresses. The model can be applied to plates made of composite materials (e.g. sandwiches) and to relatively thick plates. The Mindlin theory is a transverse shear deformable theory.

2.2   Large deflections

For large deformations, we have two special cases which are important for engineering praxis: the membrane model and the von Kármán model. The basic

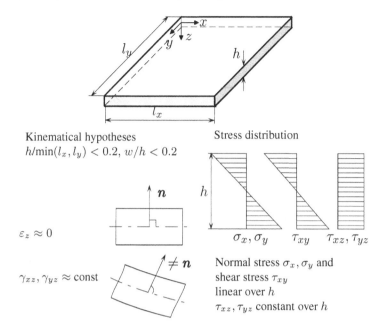

Kinematical hypotheses
$h/\min(l_x, l_y) < 0.2$, $w/h < 0.2$

Stress distribution

$\varepsilon_z \approx 0$

$\gamma_{xz}, \gamma_{yz} \approx \text{const}$

σ_x, σ_y τ_{xy} τ_{xz}, τ_{yz}

Normal stress σ_x, σ_y and
shear stress τ_{xy}
linear over h
τ_{xz}, τ_{yz} constant over h

Figure 7. Mindlin plate – basic assumptions

assumptions for the first are given in Fig. 8; there are no changes across the thickness. This model cannot describe shear stresses. Note that the membrane model is a specific structural model, since we assume a similar behavior under tension and compression. In practice a membrane is unable to respond compression.

Another way to describe large deflections is through the von Kármán model. The basic assumptions are shown in Fig. 9. The von Kármán plate theory was introduced as a engineering theory. The possibility of deducing the basic equations from the three-dimensional non-linear continuum mechanics is still under discussion, see, e.g., Ciarlet, 1990. A possible solution is given in Meenen and Altenbach, 2001.

2.3   Kirchhoff plate

Let us discuss briefly some basic features of the classical theories. As we mentioned earlier the first set of equations was given within the framework of the Kirchhoff plate theory. The possible loading cases in plate theory are shown in Fig. 10. The basic kinematic relations of the Kirchhoff plate are illustrated in Fig. 11. In the Mindlin theory, the rotations are independent entities; in Kirchhoff theory they are derivatives of the deflection, and so have simple geometrical interpretation.

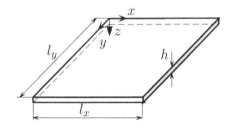

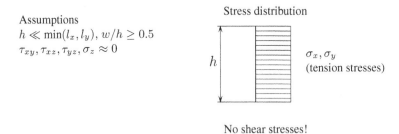

Assumptions
$h \ll \min(l_x, l_y),\ w/h \geq 0.5$
$\tau_{xy}, \tau_{xz}, \tau_{yz}, \sigma_z \approx 0$

Stress distribution

σ_x, σ_y
(tension stresses)

No shear stresses!

Figure 8.   Membrane – basic assumptions

Assumptions
$h/\min(l_x, l_y) < 0.1,$
$0.2 < w/h < 5$

Stress distribution

σ_x, σ_y τ_{xy}

shear rigid
$\varepsilon_z, \gamma_{xz}, \gamma_{yz} \approx 0$

shear rigid model
τ_{xz}, τ_{yz}

shear deformable
$\varepsilon_z \approx 0, \gamma_{xz}, \gamma_{yz} \approx$ const

shear deformable model
τ_{xz}, τ_{yz}

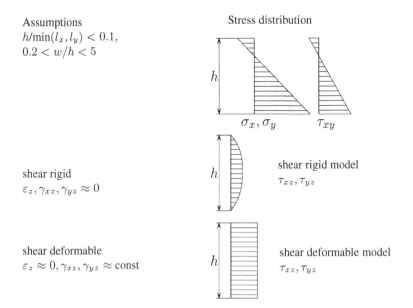

Figure 9.   Von Kármán plate – basic assumptions

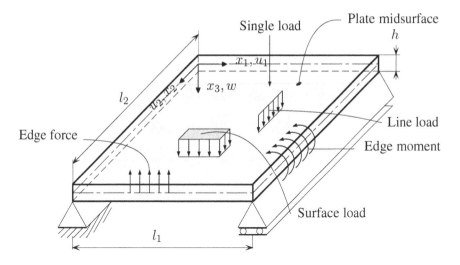

Figure 10. Loading of a plate

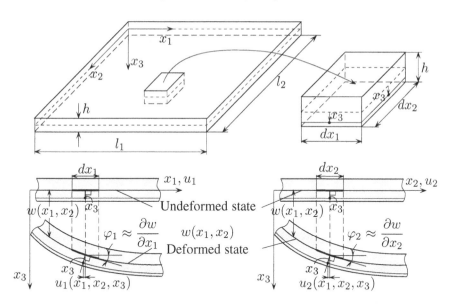

Figure 11. Kinematics of the Kirchhoff plate

In addition, in plate theory stress resultants are used instead of stresses. Such resultants are known from the strength of materials. For the Kirchhoff plate there are bending and torque moments and shear forces. They are shown in Fig. 12 for the equilibrium state.

The formulation of the boundary conditions for the Kirchhoff plate is non-trivial and widely discussed in the literature, see Altenbach et al. (1998) among

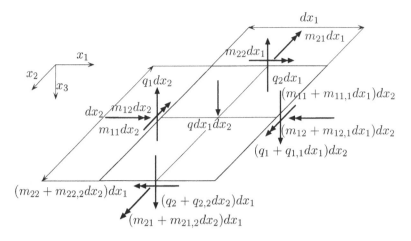

Figure 12. Stress resultants of the Kirchhoff plate

others. The reason for this is that the deflections of the Kirchhoff plate are described by partial differential equation of the fourth order, but in the general case one has to prescribe three boundary conditions at each edge. This problem is better solved in the framework of the Mindlin or Reissner theory (both are based on sixth order equation for the deflection). In addition, the Kirchhoff theory is characterized by a special description of the edge and corner forces.

Let us summarize the basic equations of the Kirchhoff theory for a rectangular plate with constant bending stiffness K (Δ is the two-dimensional Laplace operator)

- Bending equation for simply supported plate

$$K\Delta\Delta w(x_1, x_2) = q(x_1, x_2)$$

- Bending equation for elastically supported plate

$$K\Delta\Delta w(x_1, x_2) = q(x_1, x_2) - cw(x_1, x_2)$$

- Bending vibration equation for simply supported plate

$$K\Delta\Delta w(x_1, x_2, t) + \rho h\ddot{w}(x_1, x_2, t) = q(x_1, x_2, t)$$

- Bending vibration equation for elastically supported plate

$$K\Delta\Delta w(x_1, x_2, t) + \rho h\ddot{w}(x_1, x_2, t) = q(x_1, x_2, t) - cw(x_1, x_2, t)$$

In these equations ρ, h, q, c are the density of the plate material, the plate thickness, the distributed external transverse load and the Winkler foundation property, respectively. The dot denotes the time derivative. The bending stiffness can be calculated in the simplest case of constant thickness and elastic properties as $K = Eh^3/12(1 - \nu^2)$ with E as the Young's modulus and ν as the Poisson ratio. As shown in Altenbach et al. (1996) and Altenbach et al. (2004) this approach can be easily extended to laminates (**Classical Laminate Theory**).

2.4 Mindlin plate

The kinematics of the Mindlin plate differ from the Kirchoff kinematics: now two additional rotational degrees of freedom are introduced (Fig. 13). The basic equations of the Mindlin theory can be presented as follows:

$$Gh_S(\Delta w + \Phi) + q = \rho h \ddot{w},$$

$$\frac{K}{2}[(1 - \nu)\Delta\psi_1 + (1 + \nu)\Phi_{,1}] - Gh_S(\psi_1 + w_{,1}) = \frac{\rho h^3}{12}\ddot{\psi}_1,$$

$$\frac{K}{2}[(1 - \nu)\Delta\psi_2 + (1 + \nu)\Phi_{,2}] - Gh_S(\psi_2 + w_{,2}) = \frac{\rho h^3}{12}\ddot{\psi}_2$$

Here the abbreviation $\psi_{1,1} + \psi_{2,2} = \Phi$ is used. In addition, G is the shear modulus and h_S is the corrected plate thickness (see, e.g., Altenbach et al.,

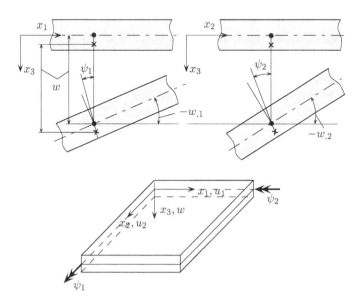

Figure 13. Kinematics of the Mindlin plate

1998). How to construct the plate thickness is still under discussion, see, for example, Altenbach (2000a,b).

After some manipulations one gets

$$\left(K\Delta - Gh_S - \frac{\rho h^3}{12} \frac{\partial^2}{\partial t^2} \right) \Phi = Gh_S \Delta w$$

and after elimination of Φ

$$\left(\Delta - \frac{\rho h}{Gh_S} \frac{\partial^2}{\partial t^2} \right) \left(K\Delta - \frac{\rho h^3}{12} \frac{\partial^2}{\partial t^2} \right) w + \rho h \frac{\partial^2 w}{\partial t^2}$$
$$= \left(1 - \frac{K}{Gh_S} \Delta + \frac{\rho h^3}{12 Gh_S} \frac{\partial^2}{\partial t^2} \right) q$$

The comparison of both models (Kirchhoff's and Mindlin's) is given in Table 1.

2.5   Von Kármán plate

Von Kármán suggested another way to extend the classical plate equations. Now the kinematics is based on displacements of different order. The in-plane displacements are small (as in Kirchhoff theory), the deflections (out-of-plane displacements) are large (see Fig. 14). As a result one gets some non-linear contributions to the components of the strain tensor. The second new idea of the von Kármán theory is the description of the equilibrium state in the deformed configuration. The stress resultants are shown in Fig. 15. Details of the von Kármán theory are given in Altenbach et al. (1998) among others.

Let us summarize the basic equations of the von Kármán theory:

- Three partial differential equations for the displacements u, v, w and

- Two partial differential equations for the deflection w and the *Airy Stress Function* Φ

For w and Φ one gets

$$K\Delta\Delta w = w_{,11}\Phi_{,22} + w_{,22}\Phi_{,11} - 2w_{,12}\Phi_{,12} + q,$$

$$\frac{1}{Eh}\Delta\Delta\Phi = -w_{,11}w_{,22} + w_{,12}^2$$

3.   Laminates and sandwiches

For laminated plates one can built up various structural models. One way is to take into account the hierarchical structure (see Fig. 16), another possibility is to model the laminate as a quasi-homogeneous system. In the latter case, one

Table 1. Comparison of Kirchhoff's and Mindlin's model (with $\Psi = \psi_{2,1} - \psi_{1,2}$)

Shear deformable plate	Shear rigid plate
Limit case $Gh_s \to \infty$ possible	

<table>
<tr>
<td>$K\Delta\Delta\tilde{w} = q, \quad \dfrac{1-\nu}{2}\dfrac{K}{Gh_s}\Delta\Psi - \Psi = 0$</td>
<td>$K\Delta\Delta w = q, \ \Psi = 0$</td>
</tr>
<tr>
<td>$w = \tilde{w} - \dfrac{K}{Gh_s}\Delta\tilde{w}$</td>
<td>$\tilde{w} = w$</td>
</tr>
<tr>
<td>$\psi_1 = -\tilde{w},_1 - \dfrac{1-\nu}{2}\dfrac{K}{Gh_s}\Psi,_2$</td>
<td>$\psi_1 = -w,_1$</td>
</tr>
<tr>
<td>$\psi_2 = -\tilde{w},_2 + \dfrac{1-\nu}{2}\dfrac{K}{Gh_s}\Psi,_1$</td>
<td>$\psi_2 = -w,_2$</td>
</tr>
<tr>
<td>$m_{11} = -K\left[\tilde{w},_{11} + \nu\tilde{w},_{22} + \dfrac{(1-\nu)^2}{2}\dfrac{K}{Gh_s}\Psi,_{12}\right]$</td>
<td>$m_{11} = -K(w,_{11} + \nu w,_{22})$</td>
</tr>
<tr>
<td>$m_{22} = -K\left[\tilde{w},_{22} + \nu\tilde{w},_{11} + \dfrac{(1-\nu)^2}{2}\dfrac{K}{Gh_s}\Psi,_{12}\right]$</td>
<td>$m_{22} = -K(w,_{22} + \nu w,_{11})$</td>
</tr>
<tr>
<td>$m_{12} = -K\left[(1-\nu)\tilde{w},_{12} - \dfrac{(1-\nu)^2}{4}\dfrac{K}{Gh_s}(\Psi,_{11} + \Psi,_{22})\right]$</td>
<td>$m_{12} = -K(1-\nu)w,_{12}$</td>
</tr>
</table>

Limit case cannot be performed	

<table>
<tr>
<td>$q_1 = -K\left[(\Delta\tilde{w}),_1 + \dfrac{1-\nu}{2}\Psi,_2\right]$</td>
<td>$q_1 = -K(\Delta w),_1$</td>
</tr>
<tr>
<td>$q_2 = -K\left[(\Delta\tilde{w}),_2 + \dfrac{1-\nu}{2}\Psi,_1\right]$</td>
<td>$q_2 = -K(\Delta w),_2$</td>
</tr>
</table>

can describe the laminate behavior by the same equations as in the previous chapters, except that the constant stiffness properties are replaced by effective stiffness properties. Sandwich plates with various cores (Fig. 17) are particular cases.

Any laminate and sandwich analysis can be based on the following definitions/statements (see, e.g., Altenbach et al., 2004):

- Laminated and sandwich plates are sequences of layers which can be modeled in a unified manner. Laminated plates consist of a large number of thin individual layers which can have various stiffness and strength properties. For the modeling of a laminated plate it will be assumed that the material behavior of each single layers is known.

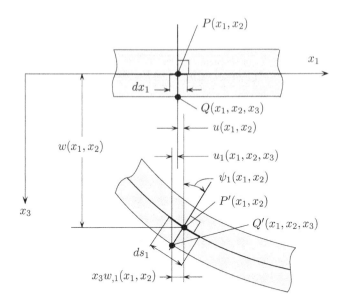

Figure 14. Kinematics of the von Kármán plate

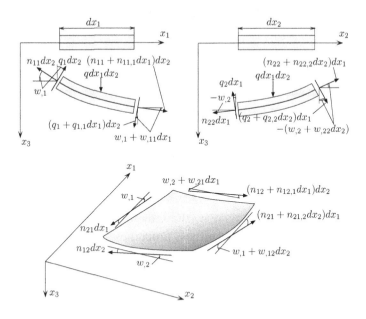

Figure 15. Stress resultants of the von Kármán plate

- The response of a laminate depends not only on the material behavior of the single layers, but also from the stacking sequence. There are symmetric, antisymmetric and unsymmetric sequences. Symmetric laminates

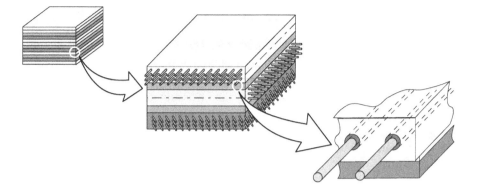

Figure 16. Hierarchical structure of UD laminated plate

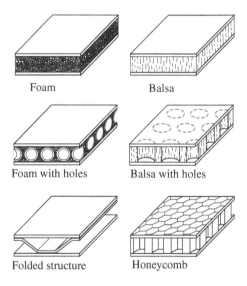

Foam

Balsa

Foam with holes

Balsa with holes

Folded structure

Honeycomb

Figure 17. Sandwich plates with various cores

lead to decoupled plane stress/bending equations; unsymmetric laminates yield coupled equations.

- Very thin laminated plates can be analyzed by the Kirchhoff theory (classical laminate theory – CLP). Depending on the laminate sequence, the
structural behavior can be isotropic, orthotropic or anisotropic. If the
transverse shear stiffness is small, laminated plates should be analyzed
by Mindlin theory (first order shear deformable theory – FOSDT).

- The exact analysis of edge or corner effects requires a three-dimensional
modeling, or the use of refined approaches (Altenbach and Becker, 2003).

- A sandwich plate is a special kind of laminated plate, consisting of two thin face sheets and one thick core. The face sheets are made from stiff materials (e.g. reinforced composites) and the core from light-weight material. The simplest model of a sandwich plate is a three-layer laminated plate with special assumptions for the face sheets and the core. For the face sheets, a plane stress state can be assumed, the core can transfer only transverse shear forces.

Finally one can draw the following conclusions (Altenbach et al., 2004):

- For laminated and sandwich plates, one has to use the known elastic properties of the single layers to deduce the effective elastic properties of the laminate.

- The deflections, the vibration behavior and the stability of these laminated elements can be calculated on the basis of the Kirchhoff or the Mindlin plate theories for anisotropic plates. Since the plane stress state may be coupled with the bending state, or the torsion with the bending, the plate equations are very complex, and only in some special cases can an analytical solution be found.

- If the local or the failure behavior become of interest, the analysis must be performed using layerwise approaches.

The classical laminate theory (general case) can be formally presented by the following vector–matrix equation

$$\begin{bmatrix} L_{11} & L_{12} & L_{13} \\ L_{21} & L_{22} & L_{23} \\ L_{31} & L_{32} & L_{33} \end{bmatrix} \begin{bmatrix} u \\ v \\ w \end{bmatrix} = \begin{bmatrix} 0 \\ 0 \\ p \end{bmatrix}$$

$L_{ij} = L_{ji}$ are differential operators of order 2, 3 or 4. For special laminated sequences, the coupled system of differential equations can be simplified. For symmetric cross-sections the system will be decoupled

$$\begin{bmatrix} L_{11} & L_{12} & 0 \\ L_{21} & L_{22} & 0 \\ 0 & 0 & L_{33} \end{bmatrix} \begin{bmatrix} u \\ v \\ w \end{bmatrix} = \begin{bmatrix} 0 \\ 0 \\ p \end{bmatrix}$$

If one has shear deformable systems (laminates with very weak layers or sandwiches with a very soft core) the first order shear deformation theory should be applied. In the general case, the following vector–matrix equation can be established

$$\begin{bmatrix} L_{11} & L_{12} & L_{13} & L_{14} & 0 \\ L_{21} & L_{22} & L_{23} & L_{24} & 0 \\ L_{31} & L_{32} & L_{33} & L_{34} & L_{35} \\ L_{41} & L_{42} & L_{43} & L_{44} & L_{45} \\ 0 & 0 & L_{53} & L_{54} & L_{55} \end{bmatrix} \begin{bmatrix} u \\ v \\ \psi_1 \\ \psi_2 \\ w \end{bmatrix} = \begin{bmatrix} 0 \\ 0 \\ 0 \\ 0 \\ p \end{bmatrix}$$

$L_{ij} = L_{ji}$ are differential operators of the order 1 or 2. For special sequences, the coupled system of differential equations can be simplified. For symmetric cross-sections the system will be decoupled

$$\begin{bmatrix} L_{11} & L_{12} \\ L_{21} & L_{22} \end{bmatrix} \begin{bmatrix} u \\ v \end{bmatrix} = \begin{bmatrix} 0 \\ 0 \end{bmatrix},$$

$$\begin{bmatrix} L_{33} & L_{34} & L_{35} \\ L_{43} & L_{44} & L_{45} \\ L_{53} & L_{54} & L_{55} \end{bmatrix} \begin{bmatrix} \psi_1 \\ \psi_2 \\ w \end{bmatrix} = \begin{bmatrix} 0 \\ 0 \\ p \end{bmatrix}$$

The specification of the vector-matrix equations is given, for example, in Altenbach et al. (2004).

4. A direct approach to plate theory

4.1 Motivation

During the GAMM-Conference in Zurich (March 1967), one of the Plenary Lectures was devoted to the Cosserat continuum and its applications (Schaefer, 1967). One message was "... everyone who is thinking about the foundations of Continuum Mechanics, will attend the world of images of the Cosserat brothers". In addition, it was noted that the applications of the Cosserat model are related to different problems in mechanics, for example, to theory of the shells, plates and beams, to the theory of plasticity, etc.

One possible classification of Cosserat-type theories can be made on this basis of dimension. In this sense one has three-dimensional theories, for which the identification of the constitutive equations is not satisfying solved. It must be noted that the two-dimensional theories (shells, plates) and one-dimensional theories (beams) are more natural because the constitutive equations can be identified by a suitable approximation to the classical three-dimensional equations.

Restricting ourself by solid mechanics problems, let us give a brief overview on the history related to the Cosserat continuum. Leonard Euler was the first to introduce the moment vector as an independent quantity. Eugène and François Cosserat applied this idea to deformable solids (see, for example, Cosserat and Cosserat, 1896, 1909). The main problem in applications was that there

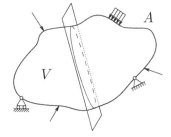

Figure 18. Continuum mechanics – basics

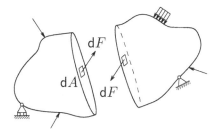

Figure 19. Non-polar continuum mechanics

were no constitutive relations. On the other hand, the advantage of this theory was that the movement of material points was characterized by independent translations and rotations (like rigid body motions). Finally, the attention of the scientists was focussed on the establishment of suitable constitutive equations for three-dimensional (see Pal'mov, 1964; Kröner, 1967; Nowacki, 1986; Rubin, 2000; Dyszlewicz, 2004; Neff, 2006), two-dimensional (see, Günther, 1961; Naghdi, 1972; Rothert and Zastrau, 1981; Altenbach, 1988; Altenbach and Zhilin, 1988; Rubin, 2000; Zhilin, 2007) and one-dimensional (Rubin, 2000; Zhilin, 2006 among others) applications.

The details of the Cosserat approach can be summarized as follows. The basic assumptions for the establishment of any solid mechanics theory are visualized in Fig. 18. Assuming the

- Cutting principle (method of sections) and

- Axiom of reciprocal action (Newton's Third Law)

we can derive some of the continuum mechanics governing equations in the traditional sense (see, e.g., Altenbach and Altenbach, 1994). The main idea for the classical continuum is shown in Fig. 19. The classical continuum theory is limited by having

- Only force actions and

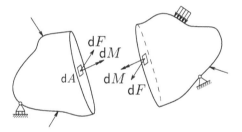

Figure 20. Polar continuum mechanics

- Only translations

These assumptions lead to in a symmetric stress tensor.

But from the basic course in Engineering Mechanics one knows that there are

- Static equilibria for the forces and the moments or

- Dynamic equilibria as balance of momentum and balance of moment of momentum

The independence/dependence of both equilibria or balances was discussed, for example, by Truesdell (1964).

There is an alternative continuum model, the so-called polar continuum (Fig. 20) based on the following assumptions Nowacki, 1986:

- Force and moment actions and

- Translations and rotations (independent quantities!)

In this case one gets in the most general case symmetric and unsymmetric stress tensors. Note that in both cases (classical and polar continua) the system of basic equations contains geometrical relations (kinematics), material independent balances of mass, momentum, moment of momentum, energy and entropy, material dependent equations (constitutive equations, evolution equations), boundary and initial conditions (Altenbach et al., 1994).

4.2 Direct approach for plates and shells

Let us give a short introduction to one of the shell theories based on the direct approach. A plate is a special case of a shell. The following assumptions will be taken into account for this variant of mechanics of two-dimensional deformable continua:

- Linear-elastic behaviour

- Geometrically non-linear theory

- Independent displacements and rotations

- Quadratic form of the strain energy

This theory is published in Zhilin (1976, 1982, 2007), Altenbach and Zhilin (1988, 2004) and Altenbach et al. (2005). A possible extension for viscoelastic material behavior was given by Altenbach, 1988.

The theory is based on the following principal definition:

> *A simple shell is a two-dimensional continuum in which the interaction between neighboring parts is due to forces and moments.*

By this definition, the preference of dynamical variables over kinematical variables is assumed.

In addition, two assumptions are taken into account:

> *The representation of the shell (homogeneous or inhomogeneous in the thickness direction) is given by a deformable surface.*

This assumption leads to the of *effective properties*, allowing one to present various classes of shells by similar equations, only the effective properties (e.g. stiffness properties) characterize the individuality of the shell under consideration.

> *Each material point of the surface is an infinitesimal body with 5 degrees of freedom (3 translations and 2 rotations).*

This assumption yields a very simple deduction of linear shell theory, but we will not use this restriction.

The kinematical model of a simple shell can be deduced by introducing two configurations. The reference configuration is related to the undeformed state

$$\{r(q^1, q^2); \; d_k(q^1, q^2)\}$$

with the position vector $r(q^1, q^2)$ and the orthonormal vectors $d_k(q^1, q^2)$. In a similar way the actual configuration (deformed state) is given by

$$\{R(q^1, q^2, t); \; D_k(q^1, q^2, t)\}, \quad D_k \cdot D_m = \delta_{km}.$$

with δ_{km} as the Kronecker symbol. The motion of the directed surface can be presented by

$$R(q, t), \quad P(q, t) \equiv D^k(q, t) \otimes d_k(q),$$

where $\otimes$ denotes the dyadic product, $P(q, t) \equiv P(q^1, q^2, t)$ denotes the rotation (turn) tensor, $\text{Det } P = +1$. The linear and angular velocities $v(q, t), \omega(q, t)$ are given as

$$v = \dot{R}, \quad \dot{P} = \omega \times P, \quad P(q^1, q^2, 0) = P_0, \quad \dot{f} \equiv \frac{\mathrm{d}f}{\mathrm{d}t}$$

The starting point for the equations of motion are the balances of momentum and of moment of momentum formulated for the two-dimensional continuum. Assuming that the Cauchy's axiom and the Gauß-Ostrogradski theorem are valid the local equations of motion can be derived. Let us formulate the first Euler equation of motion as

$$\tilde{\nabla} \cdot T + \rho F_* = \rho(v + \Theta_1^{\mathrm{T}} \cdot \omega)\dot{}, \qquad \tilde{\nabla} \equiv R^\alpha(q^1, q^2, t)\frac{\partial}{\partial q^\alpha}$$

with $T = R_\alpha \otimes T^\alpha$ as the force tensor, F_* as the mass density of the external forces, ρ, $\rho\Theta_1$ as the density and first tensor of inertia, respectively. The second Euler equation of motion can be expressed as

$$\tilde{\nabla} \cdot M + T_\times + \rho L = \rho(\Theta_1 \cdot v + \Theta_2 \cdot \omega)\dot{} + \rho v \times \Theta_1^{\mathrm{T}} \cdot \omega,$$

where $M = R_\alpha \otimes M^\alpha$ denotes the moment tensor, $T_\times \equiv R_\alpha \times T^\alpha$, L denotes the mass density of the external moments and $\rho\Theta_2$ denotes the second tensor of inertia. In the equations of motion the $\cdot$ and $\times$ denote the dot and the vector products, respectively. More details on the direct tensor calculus are given, e.g., in Naumenko and Altenbach (2007).

The equation of the balance of energy can given in the local form as

$$\rho\dot{\mathcal{U}} = T^{\mathrm{T}} \cdots \tilde{\nabla}v - T_\times \cdot \omega - M^{\mathrm{T}} \cdots \tilde{\nabla}\omega$$

with $\mathcal{U}$ - mass density of the internal energy and $(\dots)^{\mathrm{T}}$ denotes the transpose. Following Lurie, 2005, we introduce the energetic tensors

$$T_{\mathrm{e}} = (\tilde{\nabla}r)^{\mathrm{T}} \cdot T \cdot P, \quad M_{\mathrm{e}} = (\tilde{\nabla}r)^{\mathrm{T}} \cdot M \cdot P$$

and deduce another form of the balance of energy:

$$\rho\dot{\mathcal{U}} = T_{\mathrm{e}}^{\mathrm{T}} \cdots \dot{E} + M_{\mathrm{e}}^{\mathrm{T}} \cdots \dot{F}$$

Here E, F are the first and the second deformation tensors

$$E = \nabla R \cdot P - a, \quad F = (\Phi_\alpha \cdot D_k)r^\alpha \otimes d^k$$

with $\partial_\alpha P = \Phi_\alpha \times P \Rightarrow 2\Phi_\alpha = -[\partial_\alpha P \cdot P^{\mathrm{T}}]_\times$; a is the first metric tensor.

The reduced deformation tensors can be introduced as follows. Let us assume $\mathcal{U} = \mathcal{U}(E, F)$ contains 12 scalar arguments. The number of arguments can be reduced by imposing some restrictions:

- The simple shell is limited by a constant thickness.

- In addition, the linear-elastic behavior is related to non-polar materials.

In this case one can formulate

$$\boldsymbol{L} \cdot \boldsymbol{D}_3 = 0, \quad \boldsymbol{M} \cdot \boldsymbol{D}_3 = 0, \quad \boldsymbol{M}_e^{\mathrm{T}} \boldsymbol{\cdot\cdot} [(\boldsymbol{F} - \boldsymbol{b} \cdot \boldsymbol{c}) \cdot \boldsymbol{c}] + \boldsymbol{T}_e^{\mathrm{T}} \boldsymbol{\cdot\cdot} [(\boldsymbol{E} + \boldsymbol{a}) \cdot \boldsymbol{c}] = 0$$

with $\boldsymbol{b}$ as the second metric tensor, $\boldsymbol{c}$ as the discriminant tensor; and the specific energy $\mathcal{U}$ must satisfy

$$\left(\frac{\partial \mathcal{U}}{\partial \boldsymbol{E}}\right)^{\mathrm{T}} \boldsymbol{\cdot\cdot} [(\boldsymbol{E} + \boldsymbol{a}) \cdot \boldsymbol{c}] + \left(\frac{\partial \mathcal{U}}{\partial \boldsymbol{F}}\right)^{\mathrm{T}} \boldsymbol{\cdot\cdot} [(\boldsymbol{F} - \boldsymbol{b} \cdot \boldsymbol{c}) \cdot \boldsymbol{c}] = 0, \qquad \frac{\partial \rho_0 \mathcal{U}}{\partial (\boldsymbol{F} \cdot \boldsymbol{n})} = 0$$

The characteristic system of the first equation is a system of 12th order

$$\frac{\mathrm{d}}{\mathrm{d}s} \boldsymbol{E} = (\boldsymbol{E} + \boldsymbol{a}) \cdot \boldsymbol{c}, \quad \frac{\mathrm{d}}{\mathrm{d}s} \boldsymbol{F} = (\boldsymbol{F} - \boldsymbol{b} \cdot \boldsymbol{c}) \cdot \boldsymbol{c}$$

It can be shown (Courant and Hilbert, 1989) that there exist only 11 independent integrals. These independent integrals are the strain measures

$$\mathcal{E} = \frac{1}{2} \left[(\boldsymbol{E} + \boldsymbol{a}) \cdot \boldsymbol{a} \cdot (\boldsymbol{E} + \boldsymbol{a})^T - \boldsymbol{a} \right],$$

$$\boldsymbol{\Phi} = (\boldsymbol{F} - \boldsymbol{b} \cdot \boldsymbol{c}) \cdot \boldsymbol{a} \cdot (\boldsymbol{E} + \boldsymbol{a})^T + \boldsymbol{b} \cdot \boldsymbol{c} \cdot \mathcal{E} + \boldsymbol{b} \cdot \boldsymbol{c},$$

$$\boldsymbol{\gamma} = \boldsymbol{E} \cdot \boldsymbol{n},$$

$$\boldsymbol{\gamma}_* = \boldsymbol{F} \cdot \boldsymbol{n}$$

The arbitrary function $\mathcal{U}(\mathcal{E}, \boldsymbol{\Phi}, \boldsymbol{\gamma}, \boldsymbol{\gamma}_*)$ satisfies the first equation of the characteristic system. From the second equation follows that $\mathcal{U}$ does not depend on $\boldsymbol{\gamma}_*$. The tensors $\mathcal{E}, \boldsymbol{\Phi}, \boldsymbol{\gamma}$ are called the reduced deformation tensors. $\mathcal{E}$ denotes the plane tensile and shear strains, $\boldsymbol{\Phi}$ the bending and torsional strains and the $\boldsymbol{\gamma}$ transverse shear.

For a shell made from an elastic material, the strains are relatively small while the displacements and rotations can be relatively large. In such a case we can introduce the following quadratic approximation of the strain energy of simple shells:

$$2\rho_0 \mathcal{U} = 2\boldsymbol{T}_0 \boldsymbol{\cdot\cdot} \mathcal{E} + 2\boldsymbol{M}_0^T \boldsymbol{\cdot\cdot} \boldsymbol{\Phi} + 2\boldsymbol{N}_0 \cdot \boldsymbol{\gamma}$$

$$+ \mathcal{E} \boldsymbol{\cdot\cdot} {}^{(4)}\boldsymbol{C}_1 \boldsymbol{\cdot\cdot} \mathcal{E} + 2\mathcal{E} \boldsymbol{\cdot\cdot} {}^{(4)}\boldsymbol{C}_2 \boldsymbol{\cdot\cdot} \boldsymbol{\Phi} + 2\boldsymbol{\Phi} \boldsymbol{\cdot\cdot} {}^{(4)}\boldsymbol{C}_3 \boldsymbol{\cdot\cdot} \boldsymbol{\Phi}$$

$$+ \boldsymbol{\gamma} \cdot \boldsymbol{\Gamma} \cdot \boldsymbol{\gamma} + 2\boldsymbol{\gamma} \cdot ({}^{(3)}\boldsymbol{\Gamma}_1 \boldsymbol{\cdot\cdot} \mathcal{E} + {}^{(3)}\boldsymbol{\Gamma}_2 \boldsymbol{\cdot\cdot} \boldsymbol{\Phi}).$$

$\boldsymbol{T}_0, \boldsymbol{M}_0, \boldsymbol{N}_0, {}^{(4)}\boldsymbol{C}_1, {}^{(4)}\boldsymbol{C}_2, {}^{(4)}\boldsymbol{C}_3, {}^{(3)}\boldsymbol{\Gamma}_1, {}^{(3)}\boldsymbol{\Gamma}_2, \boldsymbol{\Gamma}$ are stiffness tensors of different ranks, and ρ_0 denotes the density in the reference configuration. They express the effective elastic properties of the simple shell. The differences between various classes of simple shells are connected to different expressions for the stiffness tensors. The stiffness tensors do not depend on the deformations. Thus they may be found from tests based on linear shell theory.

4.3   Tensors and their symmetry groups

After the formulation of the governing equations and the approximation of the strain energy, one open question remains – the identification of the effective properties. Various solutions of this problem are existing (Altenbach, 1991). To find the general structure of the stiffness tensors, symmetry theory must be applied. The classical theory of symmetry is not sufficient because it is valid for Euclidean tensors only. Shell theory involves non-Euclidean tensors. In addition, one has to recognize that in mechanics there are axial and polar mathematical objects. The differences are demonstrated for vectors in Fig. 21. It is easy to see that mirror reflections of an axial vector and a polar vector must be presented in different ways.

The following types of tensors are involved in shell theory:

- Polar tensors $\rho, \mathcal{U}, \mathcal{W}, \boldsymbol{u}, \dot{\boldsymbol{u}}, \boldsymbol{E}, \boldsymbol{T}, \boldsymbol{a}, \rho\boldsymbol{\Theta}_2,$ $\boldsymbol{T}_0, {}^{(4)}\boldsymbol{C}_1, {}^{(4)}\boldsymbol{C}_3, \boldsymbol{\Gamma}$

- Axial tensors $\rho\boldsymbol{\Theta}_1, \boldsymbol{\varphi}, \boldsymbol{\omega}, \boldsymbol{F}, \boldsymbol{\Phi}, \boldsymbol{b} \cdot \boldsymbol{c},$ $\boldsymbol{M}_0, {}^{(4)}\boldsymbol{C}_2$

- Polar $\boldsymbol{n}$-oriented tensors $\boldsymbol{b}, \boldsymbol{B}, \boldsymbol{\gamma}, \boldsymbol{T}_n = \boldsymbol{T} \cdot \boldsymbol{n},$ ${}^{(3)}\boldsymbol{\Gamma}_1, \boldsymbol{N}_0$

- Axial $\boldsymbol{n}$-oriented tensors $\boldsymbol{c} = -\boldsymbol{a} \times \boldsymbol{n},$ ${}^{(3)}\boldsymbol{\Gamma}_2$

The last elements on each line are the effective stiffness tensors. Let us now introduce orthogonal transformations for an arbitrary tensor of rank p as follows

$$^{(p)}\boldsymbol{S}' \equiv (\boldsymbol{n} \cdot \boldsymbol{Q} \cdot \boldsymbol{n})^{\beta} (\mathrm{Det}\boldsymbol{Q})^{\alpha} S^{i_1 \dots i_p} \boldsymbol{Q} \cdot \boldsymbol{g}_{i_1} \otimes \boldsymbol{Q} \otimes \dots \otimes \boldsymbol{Q} \cdot \boldsymbol{g}_{i_p},$$

Then one has the following: for

- Polar tensors $\alpha = 0, \beta = 0$

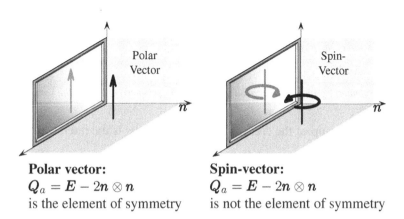

Polar vector:
$\boldsymbol{Q}_a = \boldsymbol{E} - 2\boldsymbol{n} \otimes \boldsymbol{n}$
is the element of symmetry

Spin-vector:
$\boldsymbol{Q}_a = \boldsymbol{E} - 2\boldsymbol{n} \otimes \boldsymbol{n}$
is not the element of symmetry

Figure 21.   Mirror reflections for an axial and a polar vector

- Axial tensors $\alpha = 1$, $\beta = 0$

- Polar n-oriented tensors $\alpha = 0$, $\beta = 1$

- Axial n-oriented tensors $\alpha = 1$, $\beta = 1$

Symmetries can be described in terms of the geometric operations which produce identical configurations. The set of symmetry operations and results of their combinations define a mathematical structure called a group. The symmetry operations which involve only rotations, reflections and inversion define the point group. The symmetries are described by orthogonal tensors:

- Reflection (n is the unit normal to the mirror plane)

$$Q = I - 2n \otimes n, \quad \det Q = -1,$$

- Rotation (m represents the axis and $\psi(-\pi < \psi < \pi)$ is the angle of rotation)

$$Q(\psi m) = m \otimes m + \cos \psi (I - m \otimes m) + \sin \psi m \times I, \quad \det Q = 1,$$

- Inversion

$$-I$$

How do the symmetries of the microstructure affect the physical properties? The answer comes from the Curie-Neumann principle in the physics of crystals (Nye, 1992):

> *Any type of symmetry exhibited by the point group of a crystal is possessed by every physical property of the crystal.*

or

> *For a material element and for any of its physical properties, every material symmetry transformation of the material element is a physical symmetry transformation of the physical property.*

or in other words

> *The symmetry group of the reason belongs to the symmetry group of the consequence.*

The symmetry group of the "reason" for simple shells is the intersection of:

- Symmetry of the material of the shell (fibre-reinforced material, rolled sheets)

- Symmetry of the surface shape (shell or plate) and

- Symmetry of the internal structure of the shell (laminated plates – symmetry of the layered structure with respect to the mid-surface)

For the stiffness tensors $\boldsymbol{T}_0$, $\boldsymbol{M}_0$, $\boldsymbol{N}_0$, $^{(4)}\boldsymbol{C}_1$, $^{(4)}\boldsymbol{C}_2$, $^{(4)}\boldsymbol{C}_3$, $\boldsymbol{\Gamma}$, $^3\boldsymbol{\Gamma}_1$, $^3\boldsymbol{\Gamma}_2$ there are some general constraints

$$\boldsymbol{d} \cdot\cdot \, ^{(4)}\boldsymbol{C}_1 = \, ^{(4)}\boldsymbol{C}_1 \cdot\cdot \, \boldsymbol{d}, \quad \boldsymbol{d} \cdot\cdot \, ^{(4)}\boldsymbol{C}_3 = \, ^{(4)}\boldsymbol{C}_3 \cdot\cdot \, \boldsymbol{d}, \quad \boldsymbol{c} \cdot\cdot \, \boldsymbol{\Gamma} = 0,$$

$$\boldsymbol{c} \cdot\cdot \, ^{(4)}\boldsymbol{C}_1 = \boldsymbol{0}, \quad \boldsymbol{c} \cdot\cdot \, ^{(4)}\boldsymbol{C}_2 = \boldsymbol{0}, \quad ^{(3)}\boldsymbol{\Gamma} \cdot\cdot \, \boldsymbol{c} = \boldsymbol{0}, \quad \boldsymbol{T}_0 \cdot\cdot \, \boldsymbol{c} = 0$$

$\boldsymbol{d}$ is an arbitrary tensor and $\boldsymbol{c}$ denotes an antisymmetric tensor (both of second rank).

In addition, for the solution of practical problems, one has to introduce the relations between two-dimensional and three-dimensional properties

- Forces and moments

$$\boldsymbol{T} = \langle \boldsymbol{\mu}^{-1} \cdot \boldsymbol{\sigma} \rangle, \quad \boldsymbol{M} = \langle \boldsymbol{\mu}^{-1} \cdot \boldsymbol{\sigma} \cdot \boldsymbol{c}z \rangle,$$

$\boldsymbol{\sigma}$ denotes the symmetric stress tensor of the classical theory of elasticity, $\boldsymbol{\mu} = \boldsymbol{a} - \boldsymbol{b}z$ denotes the shifter (plane symmetric tensor) and $(\ldots)^{-1}$ is the inverse of the tensor

- Displacements and rotations

$$\rho_0(\dot{\boldsymbol{u}} + \boldsymbol{\Theta}_1^T \cdot \dot{\boldsymbol{\varphi}}) = \langle \tilde{\rho}_0 \dot{\boldsymbol{u}}_* \rangle, \quad \rho_0(\boldsymbol{\Theta}_1 \cdot \dot{\boldsymbol{u}} + \boldsymbol{\Theta}_2^T \cdot \dot{\boldsymbol{\varphi}}) = \langle \tilde{\rho}_0 \dot{\boldsymbol{u}}_* \cdot \boldsymbol{c}z \rangle$$

$\boldsymbol{u}_*$ is the three-dimensional displacement vector,

- External force and moment

$$\rho_0 \boldsymbol{F}_* = \langle \tilde{\rho}_0 \tilde{\boldsymbol{F}} \rangle + \mu^+ \boldsymbol{\sigma}_n^+ + \mu^- \boldsymbol{\sigma}_n^-,$$
$$\rho_0 \boldsymbol{L} = \boldsymbol{n} \times \langle \tilde{\rho}_0 \tilde{\boldsymbol{F}}z \rangle + (h/2)\boldsymbol{n} \times (\mu^+ \boldsymbol{\sigma}_n^+ - \mu^- \boldsymbol{\sigma}_n^-)$$

$\mu^{+(-)} = 1 - (+)hH + (h^2/4)G$, H, G are the averaged and the Gaussian curvature, $\boldsymbol{\sigma}_n^{+(-)}$ are stress vectors on the upper and lower face surfaces of the shell and

$$<\ldots> = \int_{-\frac{h}{2}}^{\frac{h}{2}} \ldots dz$$

Let us discuss the local symmetry groups (LGS) of simple shells. They are given as a set of orthogonal solutions of the following system

$$\otimes_1^4 \boldsymbol{Q} \cdot \boldsymbol{C}_1 = \boldsymbol{C}_1, \quad (\text{Det})\boldsymbol{Q} \otimes_1^4 \boldsymbol{Q} \cdot \boldsymbol{C}_2 = \boldsymbol{C}_2, \quad \otimes_1^4 \boldsymbol{Q} \cdot \boldsymbol{C}_3 = \boldsymbol{C}_3,$$
$$\boldsymbol{Q} \cdot \boldsymbol{\Gamma} \cdot \boldsymbol{Q}^T = \boldsymbol{\Gamma}, \quad (\boldsymbol{n} \cdot \boldsymbol{Q} \cdot \boldsymbol{n}) \otimes_1^3 \boldsymbol{Q} \cdot \boldsymbol{\Gamma}_1 = \boldsymbol{\Gamma}_1, \quad (\boldsymbol{n} \cdot \boldsymbol{Q} \cdot \boldsymbol{n})(\text{Det}\boldsymbol{Q}) \otimes_1^3 \boldsymbol{Q} \cdot \boldsymbol{\Gamma}_2 = \boldsymbol{\Gamma}_2$$

The stiffness tensors are related to the LGS of the shell. For the surface the LGS is presented by

$$\boldsymbol{Q} \cdot \boldsymbol{a} \cdot \boldsymbol{Q}^T = \boldsymbol{a}, \quad (\boldsymbol{n} \cdot \boldsymbol{Q} \cdot \boldsymbol{n}) \boldsymbol{Q} \cdot \boldsymbol{b} \cdot \boldsymbol{Q}^T = \boldsymbol{b}$$

In the general case only three irreducible elements exist

$$1, \quad \boldsymbol{n} \otimes \boldsymbol{n} - \boldsymbol{e}_1 \otimes \boldsymbol{e}_1 + \boldsymbol{e}_2 \otimes \boldsymbol{e}_2, \quad \boldsymbol{n} \otimes \boldsymbol{n} + \boldsymbol{e}_1 \otimes \boldsymbol{e}_1 - \boldsymbol{e}_2 \otimes \boldsymbol{e}_2$$

For plates one has ($\boldsymbol{b} = \boldsymbol{0}$) and the LGS is richer. Further simplifications of the structure of the stiffness tensors can be made with additional assumptions. *But up to here we have not used the fact that the shell has a small thickness.* Note that the thinness hypotheses is the base of any traditional shell theory (see Başar and Krätzig, 1985).

Another helpful tool in the formulation of a shell theory is dimension analysis. Let us consider the special case of shell material which behaves transversely isotropically. Then the stiffness tensors depend on the following:

- $h(q)$ – the thickness of the shell

- $E(q), \nu(q)$ – the isotropic elastic properties ($G = E/[2(1 + \nu)]$)

- $\boldsymbol{a}, \boldsymbol{b}$ – the first and the second metric tensors

The following representation is valid:

$$\boldsymbol{C}_1 = \frac{Eh}{12(1 - \nu^2)} \boldsymbol{C}_1^*(h\boldsymbol{b} \cdot \boldsymbol{c}, \nu),$$

$$\boldsymbol{C}_2 = \frac{Eh^2}{12(1 - \nu^2)} \boldsymbol{C}_2^*(h\boldsymbol{b} \cdot \boldsymbol{c}, \nu),$$

$$\boldsymbol{C}_3 = \frac{Eh^3}{12(1 - \nu^2)} \boldsymbol{C}_3^*(h\boldsymbol{b} \cdot \boldsymbol{c}, \nu),$$

$$\boldsymbol{\Gamma} = Gh\boldsymbol{\Gamma}^*(h\boldsymbol{b} \cdot \boldsymbol{c}, \nu)$$

If we are now applying the thinness hypothesis

$$\|h\boldsymbol{b} \cdot \boldsymbol{c}\|^2 = (h\boldsymbol{b} \cdot \boldsymbol{c}) \cdot \cdot (h\boldsymbol{b} \cdot \boldsymbol{c})^T = \frac{h^2}{R_1^2} + \frac{h^2}{R_2^2} \ll 1,$$

we can use the following representations

$$\boldsymbol{C}_s = \frac{Eh^s}{12(1 - \nu^2)} \left[\boldsymbol{C}_s^{(0)} + \boldsymbol{C}_s^{(1)} \cdot \cdot (h\boldsymbol{b} \cdot \boldsymbol{c}) \right.$$

$$\left. + (h\boldsymbol{b} \cdot \boldsymbol{c}) \cdot \cdot \boldsymbol{C}_s^{(2)} \cdot \cdot (h\boldsymbol{b} \cdot \boldsymbol{c}) + 0(h^3) \right],$$

$$\boldsymbol{\Gamma} = Gh \left[\boldsymbol{\Gamma}^{(0)} + \boldsymbol{\Gamma}^{(1)} \cdot \cdot (h\boldsymbol{b} \cdot \boldsymbol{c}) + 0(h^2) \right]$$

with $0(h^p) \equiv 0(\|h\boldsymbol{b} \cdot \boldsymbol{c}\|^p)$, $s = 1, 2, 3$. In what follows we ignore the terms of order $0(h^2)$ with respect to 1. Only in some situations it is necessary to take into account higher order terms, e.g. for the positive definiteness of the deformation energy. Instead of the tensors $\boldsymbol{C}_i$ and $\boldsymbol{\Gamma}$ we have to consider the tensors $\boldsymbol{C}_i^{(p)}$ and $\boldsymbol{\Gamma}^{(p)}$.

It seems that the representations for the stiffness tensors do not simplify our discussion. But the $\boldsymbol{C}_i^{(p)}$ do not depend on the geometrical shape of the surface. In this case the group of symmetry of the tensors $\boldsymbol{C}_i^{(p)}$ does not include the group of symmetry (GS) of the second metric tensor $\boldsymbol{b}$.

Let us assume that the shell is made from transversely isotropic material. If $\boldsymbol{n}$ is the axis of isotropy, and the structure of the shell is such that for $\boldsymbol{b} \to \boldsymbol{0}$ we have, instead of the shell, a plate with a reference (middle) surface which is a surface of symmetry. In this case the tensors $\boldsymbol{C}_i^{(k)}$, $\boldsymbol{\Gamma}^{(k)}$ (not the tensors $\boldsymbol{C}_i, \boldsymbol{\Gamma}$) contains the following elements of symmetry

$$\boldsymbol{Q} = \pm \boldsymbol{n} \otimes \boldsymbol{n} + \boldsymbol{q}, \quad \boldsymbol{q} \cdot \boldsymbol{q}^T = \boldsymbol{a}, \quad \boldsymbol{q} \cdot \boldsymbol{n} = \boldsymbol{n} \cdot \boldsymbol{q} = \boldsymbol{0}.$$

In such a case it is easy to show that

$$\boldsymbol{C}_1^{(1)} = \boldsymbol{0}, \quad \boldsymbol{C}_2^{(0)} = \boldsymbol{0}, \quad \boldsymbol{C}_2^{(2)} = \boldsymbol{0}, \quad \boldsymbol{C}_3^{(1)} = \boldsymbol{0}, \quad \boldsymbol{\Gamma}^{(1)} = \boldsymbol{0}$$

Finally one gets the following representation of the stiffness tensors ($\boldsymbol{C}_1, \boldsymbol{C}_3$ and $\boldsymbol{\Gamma}$ from plate tests, $\boldsymbol{C}_2$ only from shell tests)

$$\boldsymbol{C}_1 = \frac{Eh}{1 - \nu^2} \left[A_1 \boldsymbol{a} \otimes \boldsymbol{a} + A_2 (\boldsymbol{r}^\alpha \otimes \boldsymbol{r}^\beta \otimes \boldsymbol{r}_\alpha \otimes \boldsymbol{r}_\beta + \boldsymbol{r}^\alpha \otimes \boldsymbol{a} \otimes \boldsymbol{r}_\alpha - \boldsymbol{a} \otimes \boldsymbol{a}) \right],$$

$$\boldsymbol{C}_2 = \frac{Eh^2}{12(1 - \nu^2)} \left[B_1 hH \, \boldsymbol{a} \otimes \boldsymbol{c} + B_2 hH (\boldsymbol{r}^\alpha \otimes \boldsymbol{c} \otimes \boldsymbol{r}_\alpha + c^{\alpha\beta} \boldsymbol{r}_\alpha \otimes \boldsymbol{a} \otimes \boldsymbol{r}_\beta) \right. $$
$$\left. + B_3 \boldsymbol{a} \otimes (h\boldsymbol{b} \cdot \boldsymbol{c} - hH\boldsymbol{c}) + B_4 h(\boldsymbol{b} \cdot \boldsymbol{c} - H\boldsymbol{c}) \otimes \boldsymbol{a} + B_5 h(\boldsymbol{b} - H\boldsymbol{a}) \otimes \boldsymbol{c} \right],$$

$$\boldsymbol{C}_3 = \frac{Eh^3}{12(1 - \nu^2)} \left[C_1 \boldsymbol{c} \otimes \boldsymbol{c} \right. $$
$$\left. + C_2 (\boldsymbol{r}^\alpha \otimes \boldsymbol{r}^\beta \otimes \boldsymbol{r}_\alpha \otimes \boldsymbol{r}_\beta + \boldsymbol{r}^\alpha \otimes \boldsymbol{a} \otimes \boldsymbol{r}_\alpha - \boldsymbol{a} \otimes \boldsymbol{a}) + h^2 H_1^2 C_4 \boldsymbol{a} \otimes \boldsymbol{a} \right],$$

$$\boldsymbol{\Gamma} = Gh\Gamma_0 \boldsymbol{a}$$

with $2H_1 = -(1/R_1) + (1/R_2)$. All stiffness expansions have the same error of $0(h^2)$.

The solution of test problems can be given as follows. The moduli A_1, A_2, $C_1, C_2, C_4, \Gamma_0, B_1, \ldots, B_5$ depend only on the Poisson's ratio. Making use of the solutions of some test problems one can obtain the following elastic moduli

$$A_1 = C_1 = \frac{1+\nu}{2}, \quad A_2 = C_2 = \frac{1-\nu}{2}, \quad \Gamma_0 = \frac{\pi^2}{12}, \quad C_4 = \frac{1-\nu}{24},$$

$$B_1 = \frac{\nu(1+\nu)}{2(1-\nu)}, \quad B_2 = 0, \quad B_3 = \frac{1+\nu}{2}, \quad B_4 = -\frac{1-\nu}{4}, \quad B_5 = -\frac{1}{2}$$

All moduli, except C_4, were found from problems in which they determine the leading terms of asymptotic expansions. Modulus C_4 is needed for the deformation energy to be positive definite. For the coefficient of transverse shear, Γ_0, the inequality $\pi^2/12 \leq \Gamma_0 < 1$ must be valid in all cases. If we are not interested in high frequencies, then it would be better to accept $\Gamma_0 = 5/(6-\nu)$. In such a case the low frequency behavior can be found more exactly. Finally, the tensors of *initial stress* T_0, M_0, Q_0 are defined by the expressions

$$T_0 = \frac{\nu h}{2(1-\nu)}a\left[n \cdot (\sigma_n^+ - \sigma_n^-)\right],$$

$$M_0 = \frac{\nu h^2}{12(1-\nu)}c\left[n \cdot (\sigma_n^+ - \sigma_n^-)\right],$$

$$Q_0 = h(1-\Gamma_0)a\left[n \cdot (\sigma_n^+ - \sigma_n^-)\right].$$

If the material of the plate or shell behaves orthotropically, one has to introduce orthotropic stiffness tensors. Their representation is

$$A = A_{11}a_1a_1 + A_{12}(a_1a_2 + a_2a_1) + A_{22}a_2a_2 + A_{44}a_4a_4,$$
$$B = B_{13}a_1a_3 + B_{14}a_1a_4 + B_{23}a_2a_3 + B_{24}a_2a_4 + B_{42}a_4a_2,$$
$$C = C_{22}a_2a_2 + C_{33}a_3a_3 + C_{34}(a_3a_4 + a_4a_3) + C_{44}a_4a_4,$$
$$\Gamma = \Gamma_1a_1 + \Gamma_2a_2, \quad \Gamma_1 = 0, \quad \Gamma_2 = 0$$

with

$$a_1 = a = e_1e_1 + e_2e_2, \quad a_2 = e_1e_1 - e_2e_2,$$
$$a_3 = c = e_1e_2 - e_2e_1, \quad a_4 = e_1e_2 + e_2e_1$$

e_1, e_2 are unit basis vectors with the orthogonality condition $a_i(i = 1, 2, 3, 4)$

$$\frac{1}{2}a_i \cdot\cdot\, a_j = \delta_{ij}, \quad \delta_{ij} = \begin{cases} 1 & i = j, \\ 0 & i \neq j \end{cases}$$

The classical stiffness tensor components are

- The in-plane (membrane) stiffness tensor components

$$A_{11} = \frac{1}{4}\left\langle \frac{E_1 + E_2 + 2E_1\nu_{21}}{1 - \nu_{12}\nu_{21}} \right\rangle,$$

$$A_{12} = \frac{1}{4} \left\langle \frac{E_1 - E_2}{1 - \nu_{12}\nu_{21}} \right\rangle,$$

$$A_{22} = \frac{1}{4} \left\langle \frac{E_1 + E_2 - 2E_1\nu_{21}}{1 - \nu_{12}\nu_{21}} \right\rangle,$$

$$A_{44} = \langle G_{12} \rangle .$$

- The mixed stiffness tensor components

$$B_{13} = -\frac{1}{4} \left\langle \frac{E_1 + E_2 + 2E_1\nu_{21}}{1 - \nu_{12}\nu_{21}} z \right\rangle,$$

$$-B_{23} = B_{14} = \frac{1}{4} \left\langle \frac{E_1 - E_2}{1 - \nu_{12}\nu_{21}} z \right\rangle,$$

$$B_{24} = \frac{1}{4} \left\langle \frac{E_1 + E_2 - 2E_1\nu_{21}}{1 - \nu_{12}\nu_{21}} z \right\rangle,$$

$$B_{42} = -\langle G_{12}z \rangle .$$

- The out-of-plane (plate) stiffness tensor components

$$C_{33} = \frac{1}{4} \left\langle \frac{E_1 + E_2 + 2E_1\nu_{21}}{1 - \nu_{12}\nu_{21}} z^2 \right\rangle,$$

$$C_{34} = -\frac{1}{4} \left\langle \frac{E_1 - E_2}{1 - \nu_{12}\nu_{21}} z^2 \right\rangle,$$

$$C_{44} = \frac{1}{4} \left\langle \frac{E_1 + E_2 - 2E_1\nu_{21}}{1 - \nu_{12}\nu_{21}} z^2 \right\rangle,$$

$$C_{22} = \langle G_{12}z^2 \rangle .$$

Details of the derivation are given in Altenbach (1987) and Altenbach (2000a). From these results we make the following conclusions:

- Analytical solutions are possible, if the distributions of the mechanical properties over the thickness are given

- Discrete layered systems – instead of integrals, sums of the weighted laminae stiffness properties have to be applied

- Functionally graded plate materials can be modeled if the distribution laws are known

Finally we estimate the nonclassical stiffness tensor components. The transverse shear stiffness are

$$\Gamma_1 = \frac{1}{2}(\lambda^2 + \eta^2)\frac{A_{44}C_{22} - B_{42}^2}{A_{44}}, \quad \Gamma_2 = \frac{1}{2}(\eta^2 - \lambda^2)\frac{A_{44}C_{22} - B_{42}^2}{A_{44}}$$

with

$$\eta = \sqrt{\frac{(\Gamma_1 + \Gamma_2)A_{44}}{A_{44}C_{22} - B_{42}^2}}, \quad \frac{d}{dz}\left(G_{1n}\frac{d\tilde{Z}}{dz}\right) + \eta^2 G_{12}\tilde{Z} = 0, \quad \frac{d\tilde{Z}}{dz}\Big|_{|z|=\frac{h}{2}} = 0$$

and

$$\lambda = \sqrt{\frac{(\Gamma_1 - \Gamma_2)A_{44}}{A_{44}C_{22} - B_{42}^2}}, \quad \frac{d}{dz}\left(G_{2n}\frac{dZ}{dz}\right) + \lambda_*^2 G_{12}Z = 0, \quad \frac{dZ}{dz}\Big|_{|z|=\frac{h}{2}} = 0$$

Details of the derivation are given in Altenbach (1987) and Altenbach (2000a).

5.   Final remarks

Till now the best overview of the history of the theory of plates and shells is given in Naghdi (1972). Todhunter and Pearson (1893), present the foundations of shell and plate theories. At the same time like Naghdi there were two overviews published in Russian (Grigolyuk and Kogan, 1972; Grigolyuk and Seleznev, 1973). Some personal remarks are given in Reissner (1985) and Zhilin (1992, 1995, 2007).

Since the direct approach is still under discussion, some references to examples performed by the author and his collaborators will be given in Altenbach and Shilin (1982), the nonlinear variant of Zhilin's theory was applied to the problem of a three-layer strip and its stability; in Altenbach (1985), the problem of the transverse vibrations of a homogeneous and an inhomogeneous plate is presented; and in Altenbach and Fedorov (1985), the problem of the boundary effect of an inhomogeneous shell is discussed. Other examples were published in Altenbach and Matzdorf (1988). Finally, some consequences of orthotropic multilayered structures were demonstrated in Altenbach (2000a,b). From all examples one can conclude that the direct approach in the theory of plates and shells leads to deeper understanding of some classical plate problems such as shear correction.

References

Altenbach, H. (1985). Untersuchung von Querschwingungen einschichtiger und mehrschichtiger Platten auf der Grundlage der linearen Theorie einfacher Schalen (in Russ.). *Dinamika i prochnost mashin*, 41:47–51.

Altenbach, H. (1987). Definition of elastic moduli for plates made from thickness-uneven anisotropic material. *Mech. Solids*, 22(1):135–0141.

Altenbach, H. (1988). Eine direkt formulierte lineare Theorie für viskoelastische Platten und Schalen. *Ingenieur–Archiv*, 58:215–228.

Altenbach, H. (1991). Modeling of viscoelastic behaviour of plates. In Życzkowski, M., editor, *Creep in Structures*, pages 531–537, Berlin. Springer.

Altenbach, H. (2000a). An alternative determination of transverse shear stiffnesses for sandwich and laminated plates. *Int. J. Solids Struct.*, 37(25):3503–3520.

Altenbach, H. (2000b). On the determination of transverse shear stiffnesses of orthotropic plates. *ZAMP*, 51:629–649.

Altenbach, J. and Altenbach, H. (1994). *Einführung in die Kontinuumsmechanik*. B. G. Teubner, Stuttgart.

Altenbach, H., Altenbach, J., and Kissing, W. (2004). *Mechanics of Composite Structural Elements*. Springer, Berlin.

Altenbach, H., Altenbach, J., and Naumenko, K. (1998). *Ebene Flächentragwerke. Grundlagen der Modellierung und Berechnung von Scheiben und Platten*. Springer, Berlin.

Altenbach, H., Altenbach, J., and Rikards, R. (1996). *Einführung in die Mechanik der Laminat– und Sandwichtragwerke. Modellierung und Berechnung von Balken und Platten aus Verbundwerkstoffen*. Deutscher Verlag für Grundstoffindustrie, Stuttgart.

Altenbach, H. and Becker, W., editors (2003). *Modern Trends in Composite Laminates Mechanics*, volume 448 of *CISM Courses and Lectures*. Springer, Wien.

Altenbach, H. and Fedorov, V.A. (1985). Untersuchung des Randeffektes einer in Dickenrichtung inhomogenen Kreiszylinderschale (in Russ.). *Dinamika i prochnost mashin*, 42:20–24.

Altenbach, J., Kissing, W., and Altenbach, H. (1994). *Dünnwandige Stab- und Stabschalentragwerke*. Vieweg, Braunschweig/Wiesbaden.

Altenbach, H. and Matzdorf, V. (1988). Zu einigen Anwendungen direkt formulierter Plattentheorien. *Wiss. Z. der TU Magdeburg*, 32(4):95–99.

Altenbach, H., Naumenko, K., and Zhilin, P.A. (2005). A direct approach to the formulation of constitutive equations for roads and shells. In Pietraszkiewicz, W. and Szymczak, C., editors, *Shell Structures – Theory and Application 2005*, pages 87–90, London. Taylor & Francis/Balkema.

Altenbach, H. and Shilin, P.A. (1982). Eine nichtlineare Theorie dünner Dreischichtschalen und ihre Anwendung auf die Stabilitätsuntersuchung eines dreischichtigen Streifens. *Technische Mechanik*, 3(2):23–30.

Altenbach, H. and Zhilin, P.A. (1988). A general theory of elastic simple shells (in Russ.). *Uspekhi Mekhaniki*, 11(4):107–148.

Altenbach, H. and Zhilin, P.A. (2004). The theory of simple elastic shells. In Kienzler, R., Altenbach, H., and Ott, I., editors, *Critical Review of the Theories of Plates and Shells and new Applications*, Lect. Notes Appl. Comp. Mech. 16, pages 1–12. Springer, Berlin.

Ambarcumyan, S.A. (1987). *Theory of Anisotropic Plates (in Russ.)*. Nauka, Moscow.

Başar, Y. and Krätzig, W.B. (1985). *Mechanik der Flächentragwerke*. Vieweg, Braunschweig, Wiesbaden.

Bollé, L. (1947a). Contribution au problème linéaire de flexin d'une plaque élastique. *Bull. Techn. Suisse Romande*, 73(21):281–285.

Bollé, L. (1947b). Contribution au problème linéaire de flexin d'une plaque élastique. *Bull. Techn. Suisse Romande*, 73(22):293–298.

Burton, W.S. and Noor, A.K. (1995). Assessment of computational models for sandwich panels and shells. *Comp. Methods Appl. Mech. Eng.*, 124(1–2):125–151.

Ciarlet, P.G. (1990). *Plates and Junctions in Elastic Multi-Structures: An Asymptotic Analysis*, volume 14 of *Collection Recherches en Mathématiques Appliquées*. Masson, Paris; Springer, Heidelberg.

Cosserat, E. and Cosserat, F. (1896). Sur la théorie de l'élasticité. *Ann. Toulouse*, 10:1–116.

Cosserat, E. and Cosserat, F. (1909). *Théorie des corps déformables*. Herman, Paris.

Courant, R. and Hilbert, D. (1989). *Methods of Mathematical Physics, Vol. 2. Partial Differential Equations*. Wiley Interscience Publication, New York.

Dyszlewicz, J. (2004). *Micropolar Theory of Elasticity*, volume 15 of *Lecture Notes in Applied and Computational Mechanics*. Springer, Berlin.

Girkmann, K. (1986). *Flächentragwerke*. Springer, Wien, 6. edition.

Gould, P.L. (1988). *Analysis of Shells and Plates*. Springer, New York et al.

Green, A.E., Naghdi, P.M., and Waniwright, W.L. (1965). A general theory of Cosserat surface. *Arch. Rat. Mech. Anal.*, 20:287–308.

Grigolyuk, E.I. and Kogan, A.F. (1972). Present state of the theory of multilayered shells (in Russ.). *Prikl. Mekh.*, 8(6):3–17.

Grigolyuk, E.I. and Seleznev, I.T. (1973). Nonclassical theories of vibration of beams, plates and shelles (in Russ.). In *Itogi nauki i tekhniki*, volume 5 of *Mekhanika tverdogo deformiruemogo tela*. VINITI, Moskva.

Günther, W. (1961). Analoge Systeme von Schalengleichungen. *Ingenieur–Archiv*, 30:160–188.

Hencky, H. (1947). Über die Berücksichtigung der Schubverzerrung in ebenen Platten. *Ingenieur–Archiv*, 16:72–76.

Irschik, H. (1993). On vibrations of layered beams and plates. *ZAMM*, 73:T34–T45.

Jaiani, G., Podio-Guidugli, P., editor (2008). *IUTAM Symposium on Relations of Shell, Plate, Beam and 3D Models*, Springer, Berlin.

Kączkowski, Z. (1980). *Płyty obliczenia statyczne*. Arkady, Warszawa.

Kienzler, R. (1982). Erweiterung der klassischen Schalentheorie; der Einfluß von Dickenverzerrung und Querschnittsverwölbungen. *Ingenieur-Archiv*, 52:311–322.

Kienzler, R. (2002). On the consistent plate theories. *Arch. Appl. Mech.*, 72:229–247.

Kienzler, R., Altenbach, H., and Ott, I., editors (2004). *Critical Review of the Theories of Plates and Shells, New Applications*, volume 16 of *Lect. Notes Appl. Comp. Mech.*, Springer, Berlin.

Kirchhoff, G.R. (1850). Über das Gleichgewicht und die Bewegung einer elastischen Scheibe. *Crelles Journal für die reine und angewandte Mathematik*, 40:51–88.

Kromm, A. (1953). Verallgemeinerte Theorie der Plattenstatik. *Ingenieur-Archiv*, 21:266–286.

Kröner, E., editor (1967). *Mechanics of Generalized Continua*, Berlin. Proc. of the IUTAM-Symposium on Generalized Cosserat Continuum and the Continuum Theory of Dislocations with Applications, Springer, Freudenstatt-Stuttgart.

Levinson, M. (1980). An accurate, simple theory of the statics and dynamics of elastic plates. *Mech. Res. Commun.*, 7(6):343–350.

Lo, K. H., Christensen, R. M., and Wu, E. M. (1977a). A high–order theory of plate deformation. Part I: Homogeneous plates. *Trans. ASME. J. Appl. Mech.*, 44(4):663–668.

Lo, K. H., Christensen, R. M., and Wu, E. M. (1977b). A high–order theory of plate deformation. Part II: Laminated plates. *Trans. ASME. J. Appl. Mech.*, 44(4):669–676.

Lurie, A.I. (2005). *Theory of Elasticity*. Foundations of Engineering Mechanics. Springer, Berlin.

Meenen, J. and Altenbach, H. (2001). A consistent deduction of von Kármán-type plate theories from threedimensional non-linear continuum mechanics. *Acta Mech.*, 147:1–17.

Mindlin, R.D. (1951). Influence of rotatory inertia and shear on flexural motions of isotropic, elastic plates. *Trans. ASME. J. Appl. Mech.*, 18:31–38.

Naghdi, P. (1972). The theory of plates and shells. In Flügge, S., editor, *Handbuch der Physik*, volume VIa/2, pages 425–640. Springer, Berlin, Heidelberg, New York.

Naumenko, K. and Altenbach, H. (2007). *Modeling of Creep for Structural Analysis. Foundations of Engineering Mechanics*. Springer, Berlin.

Neff, P. (2006). The cosserat couple modulus for continuous solids is zero viz the linearized cauchy-stress tensor is symmetric. *ZAMM*, 86(11):892–912.

Noor, A.K. and Burton, W.S. (1989a). Assessment of shear deformation theories for multilayered composite plates. *Appl. Mech. Rev.*, 42(1):1–13.

Noor, A.K. and Burton, W.S. (1989b). Stress and free vibration analysis of multilayered composite plates. *Comp. Struct.*, 11(3):183–204.

Noor, A.K. and Burton, W.S. (1990a). Assessment of computational models for multilayered anisotropic plates. *Comp. Struct.*, 14(3):233–265.

Noor, A.K. and Burton, W.S. (1990b). Assessment of computational models for multilayered composite shells. *Appl. Mech. Rev.*, 43(4):67–96.

Noor, A.K., Burton, W.S., and Bert, C.W. (1996). Computational models for sandwich panels and shells. *Appl. Mech. Rev.*, 49(3):155–199.

Nowacki, W. (1986). *Theory of Asymmetric Elasticity*. Pergamon, Oxford.

Nye, J.F. (1992). *Physical Properties of Crystals*. Oxford Science Publications, Oxford.

Pal'mov, V.A. (1964). Fundamental equations of the theory of asymmetric elasticity (in Russ.). *Prikladnaya Matematika i Mekhanika*, 28(6):1117.

Palmow, W.A. and Altenbach, H. (1982). Über eine Cosseratsche Theorie für elastische Platten. *Technische Mechanik*, 3(3):5–9.

Panc, V. (1975). *Theories of Elastic Plates*. Noordhoff International Publishing, Leyden.

Pietraszkiewicz, W. and Szymczak, C., editors (2005). *Shell Structures – Theory and Application*, Taylor & Francis/Balkema, London.

Preußer, G. (1984). Eine systematische Herleitung verbesserter Plattentheorien. *Ingenieur-Archiv*, 54:51–61.

Reddy, J.N. (1984). A simple higher–order theory for laminated composite plates. *Trans. ASME. J. Appl. Mech.*, 51:745–752.

Reddy, J.N. (1990). A general non-linear third order theory of plates with transverse deformations. *J. Non-linear Mech.*, 25(6):667–686.

Reddy, J.N. (1996). *Mechanics of Laminated Composite Plates: Theory and Analysis*. CRC Press, Boca Raton.

Reissner, E. (1944). On the theory of bending of elastic plates. *J. Math. Phys.*, 23:184–194.

Reissner, E. (1945). The effect of transverse shear deformation on the bending of elastic plates. *J. Appl. Mech.*, 12(11):A69–A77.

Reissner, E. (1947). On bending of elastic plates. *Quart. Appl. Math.*, 5:55–68.

Reissner, E. (1985). Reflection on the theory of elastic plates. *Appl. Mech. Rev.*, 38(11): 1453–1464.

Rothert, H. (1973). *Direkte Theorie von Linien- und Flächentragwerken bei viskoelastischen Werkstoffverhalten. Techn.-Wiss. Mitteilungen des Instituts für Konstruktiven Ingenieurbaus 73-2*, Ruhr-Universität, Bochum.

Rothert, H. and Zastrau, B. (1981). Herleitung einer Direktortheorie für Kontinua mit lokalen Krümmungseigenschaften. *ZAMM*, 61:567–581.

Rubin, M.B. (2000). *Cosserat Theories: Shells, Rods and Points*, volume 79 of *Solid Mechanics and Its Applications*. Springer, Berlin.

Schaefer, H. (1967). Das Cosserat-Kontinuum. *ZAMM*, 47(8):485–498.

Timoshenko, S. and Woinowsky Krieger, S. (1985). *Theory of Plates and Shells*. McGraw Hill, New York.

Todhunter, I. and Pearson, K. (1893). *A History of the Theory of Elasticity and of the Strength of Materials*, volume II. Saint-Venant to Lord Kelvin, Part II. University Press, Cambridge.

Touratier, M. (1991). An efficient standard plate theory. *Int. J. Eng. Sci.*, 29(8):901–916.

Truesdell, C. (1964). Die Entwicklung des Drallsatzes. *ZAMM*, 44(4/5):149–158.

Vlasov, V.Z. (1958). *Thinwalled Spatial Systems*. Gosstroiizdat, Moskwa. (in Russ.).

Woźniak, C. (2001). *Mechanik sprężystych płyt i powłok*, volume VIII of *Mechanika techniczna*. Wydawnictwo Naukowe PWN, Warszawa.

Wunderlich, W (1973). Vergleich verschiedener Approximationen der Theorie dünner Schalen (mit numerischen Beispielen). Techn.-Wiss. Mitteilungen des Instituts für Konstruktiven Ingenieurbaus 73-1, Ruhr-Universität, Bochum.

Zhilin, P.A. (1976). Mechanics of deformable directed surfaces. *Int. J. Solids Struct.*, 12:635–648.

Zhilin, P.A. (1982). Basic equations of non-classical theory of shells (in Russ.). In *Trudy LPI (Trans. Leningrad Polytechnical Institute) – Dinamika i prochnost mashin (Dynamics and strength of machines)*, number Nr. 386, pages 29–46. Leningrad Polytechnical Institute, Leningrad.

Zhilin, P.A. (1992). The view on Poisson's and Kirchhoff's theories of plates in terms of modern theory of plates (in Russ.). *Izvestiya RAN. Mekhanika tverdogo tela (Trans. Russ. Acad. Sci. Mech. Solids)*, Nr. 3:48–64.

Zhilin, P.A. (1995). On the classical theory of plates and the Kelvin-Teit transformation (in Russ.). *Izvestiya RAN. Mekhanika tverdogo tela (Trans. Russ. Acad. Sci. Mech. Solids)*, Nr. 4:133–140.

Zhilin, P.A. (2006). *Applied Mechanics. Theory of Thin Elastic Rods (in Russ.)*. St. Petersburg State Polytechnical University, St. Petersburg.

Zhilin, P.A. (2007). *Applied Mechanics. Foundations of the Theory of Shells (in Russ.)*. St. Petersburg State Polytechnical University, St. Petersburg.

NUMERICAL METHODS FOR THE MODELLING OF DEBONDING IN COMPOSITES

R. de Borst
Department of Mechanical Engineering
Eindhoven University of Technology
NL-5600 MB Eindhoven
The Netherlands &
LaMCoS, UMR CNRS 5514
I.N.S.A. de Lyon
F-69621 Villeurbanne, France
r.d.borst@tue.nl

Abstract   This monograph starts with a discussion of various phenomena in laminated composite structures that can lead to failure: matrix cracking, delamination between plies, and debonding and subsequent pull-out between fibres and the matrix material. Next, the different scales are discussed at which the effect of these nonlinearities can be analysed and the ways to couple analyses at these different length scales. From these scales – the macro, meso and micro-levels – the meso-level is normally used for the analysis of delamination, which is the focus of this monograph. At this level, the plies are modelled as continua and interface elements between them conventionally serve as the framework to model delamination and debonding. After a brief discussion of the cohesive–zone concept and its importance for the analysis of delamination, a particular finite element model for the plies is elaborated: the solid–like shell. This is followed by a derivation of interface elements. In the second part of this monograph more recent methods to numerically model delamination are discussed: meshfree methods, methods that exploits the partition–of–unity property of finite element shape functions, and discontinuous Galerkin methods. These approaches offer advantages over the more traditional approach that uses interface elements, as will be discussed in detail. From these more modern discretisation concepts the partition-of-unity approach seems the most promising for modelling debonding in composite structures, one advantage being that it can rather straightforwardly be incorporated in solid-like shell elements, thus enabling large-scale analyses of layered composite structures that take into account the possibility of debonding.

Keywords: multiscale analysis, debonding, delamination, finite element methods, interface elements, meshfree methods, partition-of-unity approach, solid-like shell elements, discontinuous Galerkin methods

R. de Borst and T. Sadowski (eds.) *Lecture Notes on Composite Materials – Current Topics and Achievements*

1.   Introduction

Failure in composites is governed by three mechanisms: matrix cracking, delamination, and fibre debonding and pull-out. Often, matrix cracking occurs first when loading a specimen. Together with the stress concentrations that occur near free edges and around holes, matrix cracks trigger delamination. Normally, delamination is defined as the separation of two plies of a laminated composite, although it has been observed that delamination not necessarily occurs exactly at the interface between two plies. For instance, in fibre-metal laminates delamination rather resembles a matrix crack in the epoxy layer near and parallel to the aluminium–epoxy interface.

An important issue when modelling physical phenomena is the proper definition of the scale at which the (failure) mechanism under consideration is modelled. This holds a fortiori for composites, since the in-plane dimensions of a laminated composite structure exceed the length scale at which delamination, matrix cracking and fibre debonding take place by one to several orders of magnitude. This complicates an efficient, accurate and meaningful analysis. Typically, the in-plane dimensions of a laminated structure are in the order of meters, while its thickness can be just a few millimeters. Each ply is then less than a millimeter thick. Since, at least in conventional finite element analysis, each ply has to be modelled separately in order to capture delamination between two plies, and since the aspect ratio of finite elements is limited if one wishes to obtain a reliable stress prediction, the maximum in-plane dimension of a three-dimensional solid element will be around one centimeter. It is obvious that the number of elements that is needed to model each layer is already big, and the total number of elements required to model the entire structure, including possible holes and stiffened areas, can easily exceed computational capabilities when the analyst wishes to simulate nonlinear phenomena, such as delamination.

The same reasoning holds when considering matrix cracking. For most laminated composites, matrix cracks reach a saturation distance, which is in the order of the ply thickness. This implies that, when this phenomenon is to be included in the analysis in a truly discrete format – that is, matrix cracks are modelled individually and not smeared out over the plane – the in-plane discretisation must even be somewhat finer than for an analysis that includes delamination only, roughly one order of magnitude.

A further refinement of the discretisation of several orders of magnitude is required when individual fibres are to be modelled with the aim to include debonding and pull-out of individual fibres. It is evident, that such a type of modelling exceeds computational capabilities even of the most powerful computers nowadays available, if the analysis would consider the entire structure.

Multiscale approaches provide a paradigm to by-pass the problems outlined above. In these methods, the various aspects of the entire structural problem are considered at different levels of observation, each of them characterised by a well-defined *length scale*. The different levels at which analyses are carried out, are connected either through *length scale transitions*, in which the structural behaviour at a given level is homogenised to arrive at mechanical properties at a next higher level (Ladevèze and Lubineau, 2002), or through (finite element) analyses which are conducted at two levels simultaneously and in which are connected by matching the boundary conditions at both levels (Feyel and Chaboche, 2000). In the former class of methods, the *Representative Volume Element (RVE)*, the volume of heterogeneous material that can be considered as representative at a given level of observation and is therefore amenable to homogenisation, plays an important role.

This monograph will not address methods for length scale transition or approaches for carrying out multi-level finite element analyses. Instead, we shall focus on so-called meso-level approaches, in which delamination is assumed to be the main degrading mechanism. For this purpose, the different levels of analysis – macro, meso and micro – are defined in the context of laminated composite structures. At the meso-level as well as at the micro-level, fracture along internal material boundaries, delamination and debonding, respectively, governs the failure behaviour. Most constitutive relations for such *interfaces* have in common that a so-called *work of separation* or *fracture energy* plays a central role. For this reason, a succinct discussion of cohesive-zone models, which are equipped with such a material parameter is included in the discussion.

Next, we shall formulate the three-dimensional finite element equations for arbitrarily large displacements gradients, but confined to small strains. Thermal and/or hygral strains will be included, since they are relevant because of the manufacturing process. Furthermore, solid-like shell elements will be introduced, which can be used to model the plies in a 3D-like manner, but allowing for much larger aspect ratios (up to 1000) than standard solid elements would allow.

The second part of this monograph discusses recent developments in numerical models for fracture: meshfree methods, partition-of-unity methods and discontinuous Galerkin methods. The second method seems the most promising. It exploits the partition-of-unity property of finite element shape functions and allows discontinuities to be inserted during a finite element analysis, either within a matrix, or, as used here, along interfaces between two materials. The concept will be elaborated for large displacement gradients, for the solid-like shell element discussed before and will be complemented by illustrative examples.

2.   Levels of observation

At the macroscopic or structural level the plies are normally modelled via a layered shell approach, where the different directions of the fibres in the layers are taken into account through an anisotropic elasticity model (Fig. 1). If this (anisotropic) elasticity model is augmented by a damage or plasticity model, degradation phenomena like matrix cracking, fibre pull-out or fibre breakage can also be taken into account, albeit in a smeared manner. At this level, the in-plane structural dimensions are the length scales that govern the boundary-value problem.

Indeed, discrete modelling of delamination, matrix cracking and debonding of fibres is not possible at this level, as also argued in the Introduction. The level below, where the ply thickness becomes the governing length scale, allows for the modelling of delamination and matrix cracking. At this meso-level the plies are modelled as continua and can either be assumed to behave linearly elastically or can be degraded according to a damage law. In the approach suggested by (Schellekens and de Borst, 1993, 1994), elastic anisotropy and curing of composites are taken into account by including possible thermal and hygral effects, but eventual damage which can evolve in the plies is lumped into the interface. This approach is reasonable as long as the energy dissipation due to processes like matrix cracking is small compared to the energy needed for delamination growth, as for mode-I delaminations and for mixed-mode delaminations where the fibres are (almost) parallel to the intralaminar cracks. If this condition is not met, the interface delamination model must be supplemented by a damage model for the ply, which has

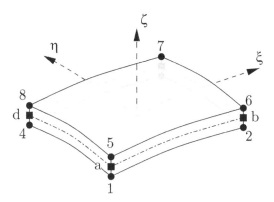

Figure 1.   Shell element for macroscopic analysis of a laminated composite structure

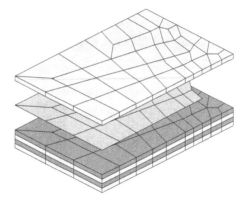

Figure 2. Finite element model of a laminated composite. The individual layers are modelled with three-dimensional, generalised plane-strain or shell elements. Interface elements equipped with a cohesive-zone model are applied between the layers

been proposed by Allix and Ladevèze (1992). A drawback of existing damage approaches for modelling intralaminar cracks, fibre breakage and debonding is that no localisation limiter is incorporated, which renders the governing equations ill-posed at a generic stage in the loading process and can result in a severe dependence of the results on the spatial discretisation (de Borst, 2004).

At the meso-level, delamination as a discrete process has conventionally been modelled as shown in Fig. 2, where the plies are considered as continua – and are discretised using standard finite elements – while the delamination is modelled in a discrete manner using special interface elements (Allix and Ladevèze, 1992; Corigliano, 1993; Schellekens and de Borst, 1993, 1994a; Allix and Corigliano 1999; Alfano and Crisfield, 2001). Generalised plane-strain elements are often used to model free-edge delamination (Schellekens and de Borst, 1993, 1994a), while stacks of solid or shell elements and interface elements are applicable to cases of delamination near holes or other cases where a three-dimensional modelling is necessary (e.g., Schipperen and Lingen, 1999).

The greatest level of detail is resolved in the analysis if the fibres are modelled individually. In such micro-level analyses the governing length scale is the fibre diameter. Possible debonding between fibre and matrix material is normally modelled via interface elements, equipped with cohesive-zone models, quite similar to models for delamination. An example is given in Fig. 3, which shows an epoxy layer, which has been reinforced uniaxially by long fibres, together with three levels of mesh refinement for a Representative Volume Element of the layer.

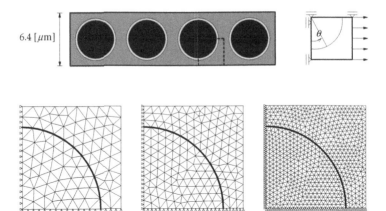

Figure 3. Layer which is unidirectionally reinforced with long fibres (above) and finite element discretisations for three different levels of refinement of a representative volume element composed of a quarter of a fibre, the surrounding epoxy matrix and the interface between fibre and epoxy (Schellekens and de Borst, 1994b)

3.   Three-dimensional framework

We denote the material coordinates of a point in the undeformed reference configuration by $\mathbf{X} = (X_1, X_2, X_3)$, while in the deformed configuration the spatial coordinates of the point become $\mathbf{x} = (x_1, x_2, x_3)$. Between $\mathbf{x}$ and $\mathbf{X}$ we have

$$\mathbf{x} = \mathbf{X} + \mathbf{u} \tag{1}$$

with $\mathbf{u}$ the displacement vector. The deformation gradient $\mathbf{F}$ is obtained by differentiating $\mathbf{x}$ with respect to $\mathbf{X}$

$$\mathbf{F} = \frac{\partial \mathbf{x}}{\partial \mathbf{X}} = \boldsymbol{i} + \frac{\partial \mathbf{u}}{\partial \mathbf{X}} \tag{2}$$

with $\boldsymbol{i}$ the second-order unit tensor. In the formulation of constitutive relations an objective strain measure is required. Because of its computational convenience the Green-Lagrange strain tensor is often selected

$$\boldsymbol{\gamma} = \frac{1}{2} \left(\frac{\partial \mathbf{u}}{\partial \mathbf{X}} + \left(\frac{\partial \mathbf{u}}{\partial \mathbf{X}} \right)^{\mathrm{T}} + \left(\frac{\partial \mathbf{u}}{\partial \mathbf{X}} \right)^{\mathrm{T}} \cdot \frac{\partial \mathbf{u}}{\partial \mathbf{X}} \right) \tag{3}$$

The incremental strain tensor is then given by

$$\Delta\boldsymbol{\gamma} = \frac{1}{2} \left(\frac{\partial \Delta\mathbf{u}}{\partial \mathbf{X}} + \left(\frac{\partial \Delta\mathbf{u}}{\partial \mathbf{X}} \right)^{\mathrm{T}} + \left(\frac{\partial \mathbf{u}}{\partial \mathbf{X}} \right)^{\mathrm{T}} \cdot \frac{\partial \Delta\mathbf{u}}{\partial \mathbf{X}} + \left(\frac{\partial \Delta\mathbf{u}}{\partial \mathbf{X}} \right)^{\mathrm{T}} \cdot \frac{\partial \mathbf{u}}{\partial \mathbf{X}} \right.$$
$$\left. + \left(\frac{\partial \Delta\mathbf{u}}{\partial \mathbf{X}} \right)^{\mathrm{T}} \cdot \frac{\partial \Delta\mathbf{u}}{\partial \mathbf{X}} \right) \tag{4}$$

The incremental strain tensor consists of a part that is linear in the incremental displacement field and a part that is quadratic in the incremental displacement field. For computational convenience we introduce

$$\Delta\gamma = \Delta\epsilon + \Delta\eta \tag{5}$$

with

$$\Delta\epsilon = \frac{1}{2}\left(\frac{\partial\Delta\mathbf{u}}{\partial\mathbf{X}} + \left(\frac{\partial\Delta\mathbf{u}}{\partial\mathbf{X}}\right)^{\mathrm{T}} + \left(\frac{\partial\mathbf{u}}{\partial\mathbf{X}}\right)^{\mathrm{T}}\cdot\frac{\partial\Delta\mathbf{u}}{\partial\mathbf{X}} + \left(\frac{\partial\Delta\mathbf{u}}{\partial\mathbf{X}}\right)^{\mathrm{T}}\cdot\frac{\partial\mathbf{u}}{\partial\mathbf{X}}\right) \tag{6}$$

$$\Delta\eta = \frac{1}{2}\left(\frac{\partial\Delta\mathbf{u}}{\partial\mathbf{X}}\right)^{\mathrm{T}}\cdot\frac{\partial\Delta\mathbf{u}}{\partial\mathbf{X}} \tag{7}$$

Furthermore, we introduce the variation of the displacement field as $\delta\mathbf{u}$, so that we can define the variation of the Green-Lagrange strain as

$$\delta\gamma = \frac{1}{2}\left(\frac{\partial\delta\mathbf{u}}{\partial\mathbf{X}} + \left(\frac{\partial\delta\mathbf{u}}{\partial\mathbf{X}}\right)^{\mathrm{T}} + \left(\frac{\partial\mathbf{u}}{\partial\mathbf{X}}\right)^{\mathrm{T}}\cdot\frac{\partial\delta\mathbf{u}}{\partial\mathbf{X}} + \left(\frac{\partial\delta\mathbf{u}}{\partial\mathbf{X}}\right)^{\mathrm{T}}\cdot\frac{\partial\mathbf{u}}{\partial\mathbf{X}}\right) \tag{8}$$

In the actual configuration and ignoring inertia effects, the balance of momentum reads

$$\nabla_{\mathbf{x}}\cdot\boldsymbol{\sigma} + \rho\mathbf{g} = \mathbf{0} \tag{9}$$

where the subscript x denotes differentiation with respect to the current configuration, $\boldsymbol{\sigma}$ is the Cauchy stress tensor, ρ is the mass density in the current configuration and $\mathbf{g}$ is the gravity acceleration. The weak form of the momentum equation is obtained in a standard manner by multiplying the balance of momentum by a test function $\mathbf{w}$ and integrating over the domain Ω. After using the divergence theorem, one obtains

$$\int_{\Omega}\nabla_{\mathbf{x}}^{\mathrm{sym}}\mathbf{w} : \boldsymbol{\sigma}\mathrm{d}\Omega = \int_{\Omega}\mathbf{w}\cdot\rho\mathbf{g}\mathrm{d}\Omega + \int_{\Gamma}\mathbf{w}\cdot\mathbf{t}\mathrm{d}\Gamma \tag{10}$$

with Γ the external boundary to the body Ω and the superscript *sym* denoting a symmetrised operator. In a total Lagrange description, which is employed predominantly in computational structural analysis, static and kinematic variables are functions of the undeformed, or reference configuration Ω^0 and it is computationally convenient to transform Eq. (10) to the reference configuration. After some algebraic manipulations, using conservation of mass, $\rho^0\mathrm{d}\Omega^0 = \rho\mathrm{d}\Omega$, and identifying the test function $\mathbf{w}$ with the variation of the displacement field, $\delta\mathbf{u}$, one obtains

$$\int_{\Omega^0}\delta\gamma : \boldsymbol{\tau}\,\mathrm{d}\Omega^0 = \int_{\Omega^0}\delta\mathbf{u}\cdot\rho^0\mathbf{g}\,\mathrm{d}\Omega^0 + \int_{\Gamma^0}\delta\mathbf{u}\cdot\mathbf{t}^0\,\mathrm{d}\Gamma^0 \tag{11}$$

where $\mathbf{t}^0$ is the (nominal) traction vector referred to the undeformed state, ρ^0 is the mass density in the undeformed configuration, Γ^0 is the surface in the undeformed state, and $\boldsymbol{\tau}$ is the second Piola-Kirchhoff stress tensor, which is related to the Cauchy stress tensor $\boldsymbol{\sigma}$ by

$$\boldsymbol{\sigma} = (\det\mathbf{F})\,\mathbf{F} \cdot \boldsymbol{\tau} \cdot \mathbf{F}^{\mathrm{T}} \tag{12}$$

with $\det\mathbf{F} = \rho/\rho^0$.

In general, Eq. (11) is highly nonlinear, because of the nonlinear dependence of $\boldsymbol{\tau}$ on $\boldsymbol{\gamma}$ and because of the nonlinear dependence of $\boldsymbol{\gamma}$ on $\mathbf{u}$: $\boldsymbol{\tau} = \boldsymbol{\tau}(\boldsymbol{\gamma}(\mathbf{u}))$. Solution of Eq. (11) is therefore achieved using some iterative procedure, usually the Newton–Raphson method in computational structural analysis. Linearising the stress-strain relation $\boldsymbol{\tau} = \boldsymbol{\tau}(\boldsymbol{\gamma})$ to give the material tangential stiffness tensor,

$$\mathbf{D} = \frac{\partial\boldsymbol{\tau}}{\partial\boldsymbol{\gamma}} \tag{13}$$

we obtain for the unknown stress $\boldsymbol{\tau}_j$ at iteration j:

$$\boldsymbol{\tau}_j = \mathbf{D} : \mathrm{d}\boldsymbol{\gamma} + \boldsymbol{\tau}_{j-1} \tag{14}$$

with $\boldsymbol{\tau}_{j-1}$ the known stress at the previous iteration $j - 1$ and the d symbol signifying the iterative change of a quantity from iteration $j - 1$ to iteration j. With Eq. (14), we obtain instead of Eq. (11):

$$\int_{\Omega^0} \delta\boldsymbol{\gamma} : \mathbf{D} : \mathrm{d}\boldsymbol{\gamma}\mathrm{d}\Omega^0 + \int_{\Omega^0} \delta\boldsymbol{\gamma} : \boldsymbol{\tau}_{j-1}\mathrm{d}\Omega^0 = \int_{\Omega^0} \delta\mathbf{u} \cdot \rho^0\mathbf{g}\mathrm{d}\Omega^0 + \int_{\Gamma^0} \delta\mathbf{u} \cdot \mathbf{t}^0\mathrm{d}\Gamma^0 \tag{15}$$

Elaborating this equation using the strain decomposition (5) and consistent linearisation leads to

$$\int_{\Omega^0} \delta\boldsymbol{\epsilon} : \mathbf{D} : \mathrm{d}\boldsymbol{\epsilon}\mathrm{d}\Omega^0 + \int_{\Omega^0} \delta\boldsymbol{\eta} : \boldsymbol{\tau}_{j-1}\mathrm{d}\Omega^0 =$$
$$\int_{\Omega^0} \delta\mathbf{u} \cdot \rho^0\mathbf{g}\mathrm{d}\Omega^0 + \int_{\Gamma^0} \delta\mathbf{u} \cdot \mathbf{t}^0\mathrm{d}\Gamma^0 - \int_{\Omega^0} \delta\boldsymbol{\epsilon} : \boldsymbol{\tau}_{j-1}\mathrm{d}\Omega^0. \tag{16}$$

After a standard discretisation of the displacement field $\mathbf{u}$, a discrete set of (nonlinear) algebraic equations is obtained.

At the beginning of each load increment, so for $j = 0$, the possible influence of hygro-thermal effects is taken into account. Assuming that there are no nonlinear effects in the plies other than hygral and thermal strains the stress increment is then given by

$$\Delta\boldsymbol{\tau} = \mathbf{D} : (\Delta\boldsymbol{\gamma} - \Delta T\boldsymbol{\alpha} - \Delta C\boldsymbol{\beta}) \tag{17}$$

with ΔT and ΔC the incremental changes in temperature and moisture content in the current loading step, respectively. The vectors $\boldsymbol{\alpha}$, $\boldsymbol{\beta}$ contain the thermal and hygroscopic expansion coefficients, respectively.

4.   Zero-thickness interface elements

The classical way to represent discontinuities in solids is to introduce zero-thickness interface elements between two neighbouring (solid) finite elements, e.g. Fig. 2 for a planar interface element. The governing kinematic quantities in interfaces are relative displacements: v_n, v_s, v_t for the normal and the two sliding modes, respectively. When collecting these relative displacements in a relative displacement vector $\mathbf{v}$, they can be related to the displacements at the upper $(+)$ and lower sides $(-)$ of the interface, $u_n^-, u_n^+, u_s^-, u_s^+, u_t^-, u_t^+$, by

$$\mathbf{v} = \mathbf{Lu} \tag{18}$$

with $\mathbf{u}^\mathrm{T} = (u_n^-, \ldots\ldots, u_t^+)$ and $\mathbf{L}$ an operator matrix:

$$\mathbf{L} = \begin{bmatrix} -1 & 0 & 0 \\ +1 & 0 & 0 \\ 0 & -1 & 0 \\ 0 & +1 & 0 \\ 0 & 0 & -1 \\ 0 & 0 & +1 \end{bmatrix} \tag{19}$$

The displacements contained in the array $\mathbf{u}$ are interpolated in a standard manner, as

$$\mathbf{u} = \mathbf{Ha} \tag{20}$$

where

$$\mathbf{H} = \mathrm{diag}\begin{bmatrix} \mathbf{h} & \mathbf{h} & \mathbf{h} & \mathbf{h} & \mathbf{h} & \mathbf{h} \end{bmatrix} \tag{21}$$

with $\mathbf{h}$ a $1 \times N$ matrix containing the interpolation polynomials, and $\mathbf{a}$ the element nodal displacement array,

$$\mathbf{a} = \left(a_n^1, \ldots\ldots, a_n^N, a_s^1, \ldots\ldots, a_s^N, a_t^1, \ldots\ldots, a_t^N \right)^\mathrm{T} \tag{22}$$

with N the total number of nodes in the interface element. The relation between nodal displacements and relative displacements for interface elements is now derived from Eqs. (18) and (20) as

$$\mathbf{v} = \mathbf{LHa} = \mathbf{B}_i \mathbf{a} \tag{23}$$

where the relative displacement-nodal displacement matrix $\mathbf{B}_i$ for the interface element reads

$$\mathbf{B}_i = \begin{bmatrix} -\mathbf{h} & \mathbf{h} & 0 & 0 & 0 & 0 \\ 0 & 0 & -\mathbf{h} & \mathbf{h} & 0 & 0 \\ 0 & 0 & 0 & 0 & -\mathbf{h} & \mathbf{h} \end{bmatrix} \tag{24}$$

For an arbitrarily oriented interface element the matrix $\mathbf{B}_i$ subsequently has to be transformed to the local coordinate system of the integration point or node-set.

For analyses of fracture propagation that exploit interface elements, cohesive-zone models (Dugdale, 1960; Barenblatt, 1962) are used almost exclusively. In this class of fracture models, a discrete relation is adopted between the interface tractions $\mathbf{t}_i$ and the relative displacements $\mathbf{v}$:

$$\mathbf{t}_i = \mathbf{t}_i(\mathbf{v}, \kappa) \tag{25}$$

with κ a history parameter. After linearisation, necessary to use a tangential stiffness matrix in an incremental-iterative solution procedure, one obtains:

$$\dot{\mathbf{t}}_i = \mathbf{T}\dot{\mathbf{v}} \tag{26}$$

with $\mathbf{T}$ the material tangent stiffness matrix of the discrete traction-separation law:

$$\mathbf{T} = \frac{\partial \mathbf{t}_i}{\partial \mathbf{v}} + \frac{\partial \mathbf{t}_i}{\partial \kappa}\frac{\partial \kappa}{\partial \mathbf{v}} \tag{27}$$

A key element is the presence of a work of separation or fracture energy, $\mathcal{G}_c$, which governs crack growth and enters the interface constitutive relation (25) in addition to the tensile strength f_t. It is defined as the work needed to create a unit area of fully developed crack:

$$\mathcal{G}_c = \int_{v_n=0}^{\infty} \sigma \mathrm{d}v_n \tag{28}$$

with σ the stress across the *fracture process zone*. It thus equals the area under the decohesion curves as shown in Fig. 4. Evidently, cohesive-zone models as defined above are equipped with an *internal length scale*, since the quotient $\mathcal{G}_c/E$, with E a stiffness modulus for the surrounding continuum, has the dimension of length.

Conventional interface elements have to be inserted in the finite element mesh at the beginning of the computation, and therefore, a finite stiffness must

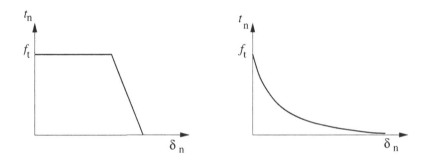

Figure 4.   Stress-displacement curves for ductile separation (left) and for quasi-brittle separation (right)

be assigned in the pre-cracking phase with at least the diagonal elements being non-zero. Prior to crack initiation, the stiffness matrix in the interface element therefore reads

$$\mathbf{T} = \begin{bmatrix} d_n & 0 & 0 \\ 0 & d_s & 0 \\ 0 & 0 & d_t \end{bmatrix} \tag{29}$$

with d_n the stiffness normal to the interface and d_s and d_t the tangential stiffnesses. With the material tangent stiffness matrix $\mathbf{T}$, the element tangent stiffness matrix can be derived in a straightforward fashion, starting from the weak form of the equilibrium equations, as

$$\mathbf{K} = \int_{\Gamma_i} \mathbf{B}_i^{\mathrm{T}} \mathbf{T} \mathbf{B}_i \mathrm{d}\Gamma \tag{30}$$

where the integration domain extends over the surface of the interface Γ_i. For comparison with methods that will be discussed in the remainder of this paper, we expand the stiffness matrix in the pre-cracking phase (cf. Schellekens and de Borst, 1992):

$$\mathbf{K} = \begin{bmatrix} \mathbf{K}_n & \mathbf{0} & \mathbf{0} \\ \mathbf{0} & \mathbf{K}_s & \mathbf{0} \\ \mathbf{0} & \mathbf{0} & \mathbf{K}_t \end{bmatrix} \tag{31}$$

with the submatrices $\mathbf{K}_\pi$, $\pi = n, s, t$ defined as

$$\mathbf{K}_\pi = d_\pi \begin{bmatrix} \mathbf{h}^{\mathrm{T}}\mathbf{h} & -\mathbf{h}^{\mathrm{T}}\mathbf{h} \\ -\mathbf{h}^{\mathrm{T}}\mathbf{h} & \mathbf{h}^{\mathrm{T}}\mathbf{h} \end{bmatrix} \tag{32}$$

with d_π the (dummy) stiffnesses in the interface prior to crack initiation.

An example where the potential of cohesive-zone models can be exploited fully using conventional discrete interface elements, is the analysis of delamination in layered composite materials (Allix and Ladevèze, 1992; Schellekens and de Borst, 1993, 1994a). Since the propagation of delaminations is restricted to the interfaces between the plies, inserting interface elements at these locations permits an exact simulation of the failure mode.

Due to mismatch of the Poisson effect between the layers of a laminated structure, caused by the different orientation of the fibres, interlaminar stresses will develop between the plies at the free edges. At a generic stage in the loading process, these edge stresses will lead to delamination. Depending on the stacking sequence of the laminate and the position of the delamination zone in the laminate, delamination occurs purely as mode-I delamination or as delamination due to a combination of several cracking modes, so-called mixed-mode delamination. For the three-dimensional example of Fig. 5, we will consider a lay-up that causes pure mode-I delamination, which is the dominant mode if

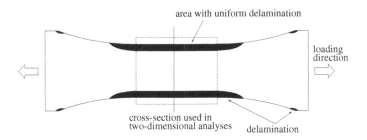

Figure 5. T-bone shaped AS-3501-06 graphite-epoxy laminated strip subjected to uniaxial loading

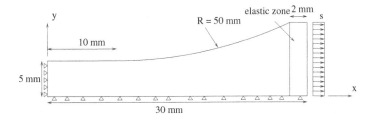

Figure 6. Quarter of T-bone shaped laminated strip

delamination occurs in the mid-plane of a symmetric laminate. Consequently, only the upper (or equivalently, the lower) half of the laminate needs to be analysed. The interface delamination model was based on a damage formalism, see Schipperen and Lingen (1999) for details.

The strip that has been analysed, has a laminate lay-up of $[25, -25, 90]_s$ and is manufactured of an AS-3501-06 graphite-epoxy. The specimen that has been analysed is depicted in Fig. 6 in more detail. The linear elastic ends of the specimen are a simplification of the real situation in an experiment and have been included in the analyses to limit the influence of the boundary conditions. Furthermore, to reduce the computation time, the radius of the transition zone has been taken fairly small compared to data suggested in norms. The COD versus axial stress, measured as the average stress in the narrow part of the strip, is shown in Fig. 7.

Numerical solutions of boundary-value problems involving materials that show a descending branch after reaching a peak load level, can be highly mesh sensitive, e.g., de Borst (2004). However, in the present situation, where the degrading phenomena are limited to a discrete interface where the crack opening is controlled by a fracture energy (cohesive-zone approach), the boundary value problem remains well-posed and, consequently, no mesh sensitivity should be observed. This is confirmed in a mesh refinement study of a three-dimensional rectangular plate, Fig. 8, which is used to approximate the original

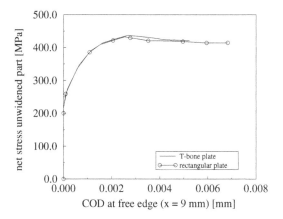

Figure 7. COD *vs* axial stress for the full T-bone specimen and for an approximated 3D solution using the rectangular specimen of Fig. 8

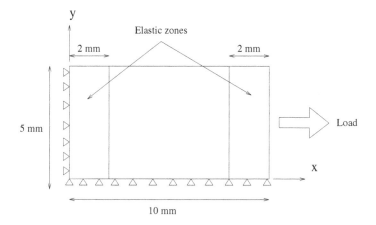

Figure 8. Quarter of the rectangular specimen used in the approximate 3D solutions

T-bone specimen, but, because of its simpler geometry, is less expensive in mesh refinement studies. The load-displacement curves for the original T-bone specimen and the approximate 3D specimen are close, Fig. 7, justifying the approximation for the purpose of a mesh refinement study.

Three different meshes have been used in the calculations. The coarse mesh consisted of 20 elements over the width and 25 elements over the length of the plate. For the two finer meshes the element distribution over the width was not equidistant. For the 2.5 mm of the width of the plate closest to the free edge a finer mesh was used. This leads to 35 elements over the width and 25 elements over the length of the plate for the second mesh and to 70 elements over the width and 50 elements over the length of the place for the finest mesh.

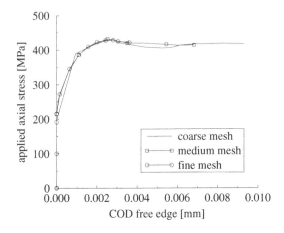

Figure 9.   Mesh sensitivity studies for the 3D rectangular specimen

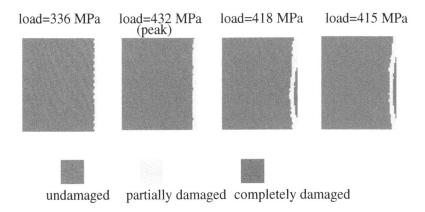

Figure 10.   Evolution of the delamination zone in simplified three-dimensional analysis

The crack opening displacement of a node near the centre of the free edge has been plotted versus the applied axial stress for all three meshes in Fig. 9. No mesh sensitivity can be noticed. In Fig. 10 the delamination zone of the plate is shown at several stages during the computation. Until the peak load the delamination is uniform, since the slight waviness is purely due to visualisation aspects. However, in the descending branch of Fig. 9 the delamination zone becomes more and more non-uniform.

Figure 11 shows an example of a uniaxially loaded laminate that fails under mixed-mode loading. Experimental and numerical results (which were obtained *before* the tests were carried out) show an excellent agreement, Fig. 11, which gives the ultimate strain of the sample for different numbers of plies in the laminate (Schellekens and de Borst, 1994a). A clear thickness (size) effect is

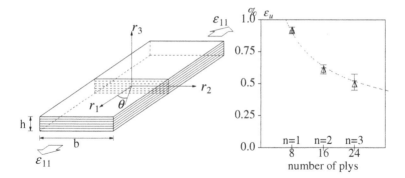

Figure 11. Left: Uniaxially loaded laminated strip. Right: Computed and experimentally determined values for the ultimate strain ϵ_u as a function of the number of plies (Schellekens and de Borst, 1994a). Results are shown for laminates consisting of eight plies ($n = 1$), sixteen plies ($n = 2$) and twenty-four plies ($n = 3$). The triangles, which denote the numerical results, are well within the band of experimental results. The dashed line represents the inverse dependence of the ultimate strain on the laminate thickness

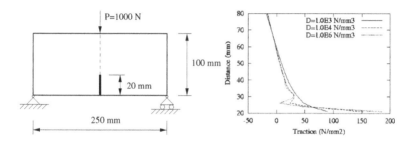

Figure 12. Left: Geometry of symmetric, notched three-point bending beam. Right: Traction profiles ahead of the notch using linear interface elements with Gauss integration. Results are shown for different values of the 'dummy' stiffness $D = d_n$ in the pre-cracking phase (Schellekens and de Borst, 1992)

obtained as a direct consequence of the inclusion of the fracture energy in the model.

As stipulated, conventional interface elements have to be inserted *a priori* in the finite element mesh. The undesired *elastic* deformations can be largely suppressed by choosing a high value for the stiffness d_n. However, the off-diagonal coupling terms of the submatrix $\mathbf{h}^T\mathbf{h}$ that enters the stiffness matrix of the interface elements, cf. Eq. (32), can lead to spurious traction oscillations in the pre-cracking phase for high stiffness values (Schellekens and de Borst, 1992). This, in turn, may cause erroneous crack patterns. An example of an oscillatory traction pattern ahead of a notch is given in Fig. 12. Moreover, when analysing dynamic fracture, spurious wave reflections can occur as a result of the introduction of such artificially high stiffness values prior to the onset of

delamination. Thirdly, the necessity to align the mesh with the potential planes
of delamination, restricts the modelling capabilities.

5.   Solid-like shell formulation

We consider the thick shell shown in Fig. 13. The position of a material
point in the shell in the undeformed configuration can be written as a function
of the three curvilinear coordinates $[\xi, \eta, \zeta]$:

$$\mathbf{X}(\xi, \eta, \zeta) = \mathbf{X}_0(\xi, \eta) + \zeta \mathbf{D}(\xi, \eta) \tag{33}$$

where $\mathbf{X}_0(\xi, \eta)$ is the projection of the point on the mid-surface of the shell
and $\mathbf{D}(\xi, \eta)$ is the thickness director in this point:

$$\mathbf{X}_0(\xi, \eta) = \frac{1}{2}\left[\mathbf{X}_t(\xi, \eta) + \mathbf{X}_b(\xi, \eta)\right] \tag{34}$$

$$\mathbf{D}(\xi, \eta) = \frac{1}{2}\left[\mathbf{X}_t(\xi, \eta) - \mathbf{X}_b(\xi, \eta)\right] \tag{35}$$

The subscripts $(\cdot)_t$ and $(\cdot)_b$ denote the projections of the variable onto the
top and bottom surface, respectively. The position of the material point in the
deformed configuration $\mathbf{x}(\xi, \eta, \zeta)$ is related to $\mathbf{X}(\xi, \eta, \zeta)$ via the displacement
field $\boldsymbol{\phi}(\xi, \eta, \zeta)$ according to

$$\mathbf{x}(\xi, \eta, \zeta) = \mathbf{X}(\xi, \eta, \zeta) + \boldsymbol{\phi}(\xi, \eta, \zeta) \tag{36}$$

where

$$\boldsymbol{\phi}(\xi, \eta, \zeta) = \mathbf{u}_0(\xi, \eta) + \zeta \mathbf{u}_1(\xi, \eta) + (1 - \zeta^2)\mathbf{u}_2(\xi, \eta) \tag{37}$$

In this relation, $\mathbf{u}_0$ and $\mathbf{u}_1$ are the displacements of $\mathbf{X}_0$ on the shell mid-surface,
and the thickness director $\mathbf{D}$, respectively:

$$\mathbf{u}_0(\xi, \eta) = \frac{1}{2}\left[\mathbf{u}_t(\xi, \eta) + \mathbf{u}_b(\xi, \eta)\right] \tag{38}$$

$$\mathbf{u}_1(\xi, \eta) = \frac{1}{2}\left[\mathbf{u}_t(\xi, \eta) - \mathbf{u}_b(\xi, \eta)\right] \tag{39}$$

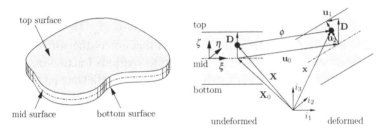

Figure 13.   Kinematic relations of the solid-like shell element

and $\mathbf{u}_2(\xi, \eta)$ denotes the internal stretching of the element, which is colinear with the thickness director in the deformed configuration and is a function of an additional 'stretch' parameter w:

$$\mathbf{u}_2(\xi, \eta) = w(\xi, \eta)[\mathbf{D} + \mathbf{u}_1(\xi, \eta)] \qquad (40)$$

The displacement field ϕ is considered as a function of two kinds of variables; the ordinary displacement field $\mathbf{u}$, which will be split in a displacement of the top and bottom surfaces $\mathbf{u}_t$ and $\mathbf{u}_b$, respectively, and the internal stretch parameter w:

$$\phi = \phi(\mathbf{u}_t, \mathbf{u}_b, w) \qquad (41)$$

The derivation of the strains and the finite element formulation can be found in Parisch (1995), or in Remmers *et al.* (2003).

Using the solid-like shell element, the behaviour of a Glare panel with a circular initial delamination and a sinusoidally shaped out-of-plane imperfection (with an amplitude of 0.003 *mm*) subject to a compressive load has been examined. The failure mechanism is slightly complicated, since the delaminated zone grows in a direction perpendicular to the main loading direction. As a result, the delaminated area transforms from a circular area into an ellipsoidal one. Consequently, the buckling mode will change as well, and some parts of the top layer will tend to move inwards. For this reason, the possibility of self-contact has been included and a contact algorithm has been activated.

The specimen of Fig. 14 consists of an aluminium layer with thickness $h_1 = 0.2$ *mm* and a Glare3 0/90° prepreg layer with a thickness $h_2 = 0.25$ *mm*. An initially circular delamination area with radius 8 *mm* is assumed. The layers are attached to a thick backing plate in order to prevent global buckling. A uniaxial compressive loading in x-direction is considered ($\sigma_x = -\sigma_0$, $\sigma_y = 0.0$).

The finite element mesh is shown in Fig. 15. The material parameters for the Glare3 layer are taken from Hashagen and de Borst (2000), see Table 1. The ultimate traction in normal direction in tension and compression are assumed to be $\bar{t}_n^t = 50$ *MPa* and $\bar{t}_n^c = 150$ *MPa*, respectively, and the ultimate traction in the two transverse directions equals $\bar{t}_{s1} = \bar{t}_{s2} = 25$ *MPa*. The work of separation is $\mathcal{G}_c = 1.1$ *N/mm*. An initial stiffness of the interface elements of $d_n = 50{,}000$ *N/mm²* has been assumed.

The analytical estimation for the local buckling load of a clamped unidirectional panel with thickness h_1 subjected to an axial compressive load σ_0 has been derived by Shivakumar and Whitcomb (1985). For this configuration, the lowest critical buckling load is equal to $\sigma_0 = 113.2$ *MPa*.

For the contact algorithm the penalty stiffness has been set equal to the initial stiffness of the interface elements with the delamination model, $d_{pen} = 50{,}000$ *MPa*. The out-of-plane displacement of the centre point of the panel

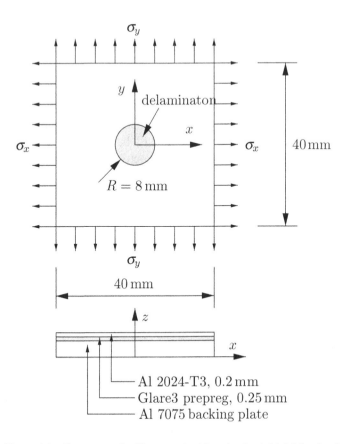

Figure 14.   Geometry of a Glare panel with a circular initial delamination

is shown in Fig. 16. The local buckling load is in agreement with an eigenvalue analysis (Remmers and de Borst, 2002). Initial delamination growth does not start until a load level $\sigma_0 = 300\ MPa$, while progressive delamination begins at an external load level $\sigma_0 \approx 950\ MPa$. As this value is far beyond normal stress levels, the analysis suggests that delamination buckling is of little concern in uniaxially compressed Glare panels. As expected, the delamination extends in a direction perpendicular to the loading direction, Fig. 17.

6.   Meshfree methods

In view of the limitations discussed above, discretisation methods have been sought for that facilitate an improved resolution in the presence of stress singularities for *crack initiation* and that obviate the need for elaborate remeshing after *crack propagation*. Nayroles *et al.* (1992) and Belytschko *et al.* (1994) have introduced interpolants that are based on the concept of moving least

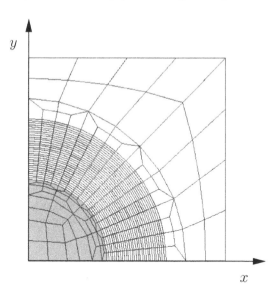

Figure 15. Mesh used for the simulation of delamination growth in the Glare panel. The initial delamination is located at the darker elements. Note that just one quarter of the panel ($x > 0$, $y > 0$) has been modelled

Table 1. Material parameters for 0/90° Glare3

E_{11}	33 170 MPa	G_{12}	5 500 MPa	ν_{12}	0.195
E_{22}	33 170 MPa	G_{23}	5 500 MPa	ν_{23}	0.032
E_{33}	9 400 MPa	G_{13}	5 500 MPa	ν_{13}	0.06

squares. In such an interpolation scheme, the approximation function $u^h(\mathbf{x})$ is expressed as the inner product of a vector $\mathbf{p}(\mathbf{x})$ and a vector $\mathbf{a}(\mathbf{x})$,

$$u^h(\mathbf{x}) = \mathbf{p}^{\mathrm{T}}(\mathbf{x})\mathbf{a}(\mathbf{x}) \tag{42}$$

in which $\mathbf{p}(\mathbf{x})$ contains basis terms that are functions of the coordinates $\mathbf{x}$. Normally, monomials such as $1, x, y, z, x^2, xy, \ldots\ldots$ are chosen, although also more sophisticated functions can be taken. The array $\mathbf{a}(\mathbf{x})$ contains the coefficients of the basis terms. In a moving least squares interpolation each node is assigned a weight function which renders the coefficients $\mathbf{a}$ non-uniform. These weight functions w_i appear in the sum J^{mls} as

$$J^{\mathrm{mls}} = \sum_{i=1}^{n} w_i(\mathbf{x}) \left(\mathbf{p}^{\mathrm{T}}(\mathbf{x}_i)\mathbf{a}(\mathbf{x}) - u_i \right)^2 \tag{43}$$

that has to be minimised with respect to $\mathbf{a}(\mathbf{x})$. Typical choices for the weight functions are Gaussian distributions, splines, or radial basis functions, whereby

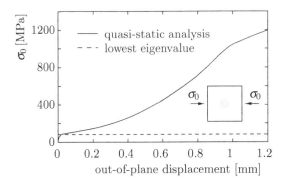

Figure 16. Out-of-plane displacement of top layer versus applied axial compressive load σ_0. The dashed line corresponds to the critical buckling load obtained by an eigenvalue analysis (Remmers and de Borst, 2002)

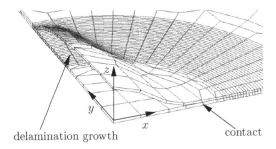

Figure 17. Final deformation of the Glare laminate under uniaxial loading (Remmers and de Borst, 2002)

the domain of influence may take the shape of a disc (sphere) or rectangle (brick) in two (three) dimensions. Elaboration of the stationarity requirement of J^{mls} with respect to $\mathbf{a}(\mathbf{x})$ gives

$$\frac{\partial J^{\mathrm{mls}}}{\partial \mathbf{a}(\mathbf{x})} = \sum_{i=1}^{n} w_i(\mathbf{x}) \left[2\mathbf{p}(\mathbf{x}_i)\mathbf{p}^{\mathrm{T}}(\mathbf{x}_i)\mathbf{a}(\mathbf{x}) - 2\mathbf{p}(\mathbf{x}_i)u_i \right] = \mathbf{0} \qquad (44)$$

Thus, $\mathbf{a}(\mathbf{x})$ can be obtained as

$$\mathbf{a}(\mathbf{x}) = \mathbf{A}^{-1}(\mathbf{x})\mathbf{C}(\mathbf{x})\mathbf{u} \qquad (45)$$

where $\mathbf{u}$ contains all u_i, and

$$\mathbf{A}(\mathbf{x}) = \sum_{i=1}^{n} w_i(\mathbf{x})\mathbf{p}(\mathbf{x}_i)\mathbf{p}^{\mathrm{T}}(\mathbf{x}_i) \qquad (46a)$$

$$\mathbf{C}(\mathbf{x}) = [w_1(\mathbf{x})\mathbf{p}(\mathbf{x}_1), w_2(\mathbf{x})\mathbf{p}(\mathbf{x}_2), \ldots, w_n(\mathbf{x})\mathbf{p}(\mathbf{x}_n)] \qquad (46b)$$

Equation (45) is substituted into Eq. (42), which leads to

$$u^h(\mathbf{x}) = \mathbf{p}^T(\mathbf{x})\mathbf{A}^{-1}(\mathbf{x})\mathbf{C}(\mathbf{x})\mathbf{u} \qquad (47)$$

and the matrix that contains the shape functions $\mathbf{H}(\mathbf{x})$ can be identified as

$$\mathbf{H}(\mathbf{x}) = \mathbf{p}^T(\mathbf{x})\mathbf{A}^{-1}(\mathbf{x})\mathbf{C}(\mathbf{x}) \qquad (48)$$

Shape functions which are generated in this manner, are usually not of a polynomial form, even though $\mathbf{p}(\mathbf{x})$ contains only polynomial terms. When moving least squares shape functions are used, the weight functions that are attached to each node determine the degree of continuity of the interpolants and the extent of the support of the node. A high degree of continuity can thus be achieved easily, so that steep stress gradients can be captured accurately, which is beneficial for the proper prediction of crack initiation. The fact that the extent of the support is determined by the weight function is in contrast with the finite element method. Consequently, there are no *elements* needed to define the support of a node. A mesh is not necessary and approximation methods based on moving least squares functions are often termed *meshfree* or *meshless methods*. However, the support of one node normally includes several other nodes and is therefore less compact than with finite element methods and, therefore, leads to a larger bandwidth of the system of equations.

Discontinuous shape functions for use in fracture mechanics applications can be obtained in a straightforward manner by truncating the appropriate weight functions. Implicitly, the same procedure is applied as for nodes close to the boundary of the domain: the part of the domain of influence that falls outside the computational domain is simply not taken into account in the integration.

A different situation arises when the crack does not pass completely through a domain of influence, so that the crack tip lies inside the support. Figure 18

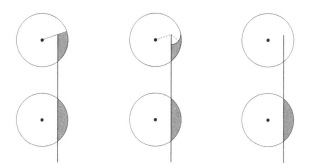

Figure 18. Domains of influence intersected by a crack or the crack tip: truncation of the weight function according to the visibility criterion (left), diffraction criterion (center) and the see-through criterion (right) — shaded areas denote the neglected part of the domain of influence

illustrates three different procedures how to truncate the domain of influence in the case of intersection by a crack, see also Fleming *et al.* (1997). In the visibility criterion the connectivity between an integration point and a node is taken into account if and only if a line can be drawn that is not intersected by a non-convex boundary. The resulting shape functions are not only discontinuous over the crack path, but also over the line that connects node and crack tip. Although convergent results can be obtained, the presence of discontinuities in the shape functions beyond the crack path is less desirable. As an alternative, it has been suggested to redefine the weight function. For instance, the line that connects node and integration point can be wrapped around the crack tip, in a similar way as light diffracts around sharp edges – hence the name diffraction criterion – or, the visibility criterion can be adapted such that some transparency is assigned to the part of the crack close to the crack tip. In either way, shape functions are obtained that are smooth and continuous for the part of the domain not intersected by the crack. In the see-through or continuous path criterion, truncation of the weight function only occurs when the domain of influence is completely intersected by the crack path. In this fashion, the effect of the crack propagation is delayed, and inaccuracies have been reported for this method (Fleming *et al.*, 1997).

Another issue is the spatial resolution around the crack path and the crack tip. For linear-elastic fracture mechanics applications, the shape functions should properly capture the $r^{-1/2}$ - singularity near the crack tip in order to accurately compute the stress intensity factors. Apart from a nodal densification around the crack tip, this can be achieved by *locally* enriching the base vector $\mathbf{p}$ through the addition of the set

$$\boldsymbol{\psi} = \left(\sqrt{r} \cos(\theta/2), \sqrt{r} \sin(\theta/2), \ \sqrt{r} \sin(\theta/2) \sin(\theta), \ \sqrt{r} \cos(\theta/2) \sin(\theta) \right)^{\mathrm{T}}$$

(49)

with r is the distance from the crack tip and θ is measured from the current direction of crack propagation. Alternatively, these functions can be added to the sum of Eq. (43). This is possible by virtue of the fact that, similar to conventional finite element shape functions, shape functions obtained from a moving least squares approximation satisfy the partition-of-unity property, an issue to which we will return in the next Section.

As an example of a meshfree simulation of crack propagation using linear elastic fracture mechanics concepts, dynamic crack extension in a three-dimensional cube is considered (Krysl and Belytschko, 1999). A penny-shaped crack is initially present, which extends internally in the cube. When the crack reaches the free surfaces of the cube, a full separation of the cube takes place. In Fig. 19 the development of the crack is plotted for eight successive stages. It shows the ability of meshfree methods to describe not only cracks as line segments, but also as faces in three-dimensional analyses.

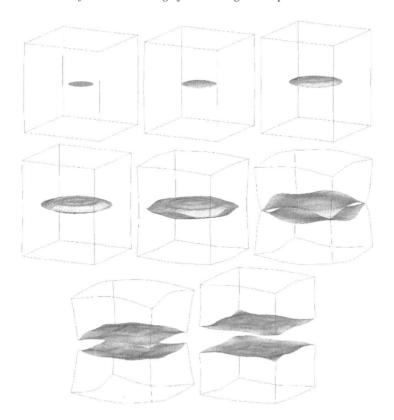

Figure 19. Cube with centered penny-shape crack – propagation of the crack towards the free surfaces of the specimen (Krysl and Belytschko, 1999)

The high degree of continuity that is incorporated in meshfree methods makes them ideally suited for localisation and failure analyses that adopt higher-order continuum models. Also, the flexibility is increased compared to conventional finite element methods, since there is no direct connectivity, which makes placing additional nodes in regions with high strain gradients particularly simple. An example is offered in Fig. 20 for a fourth-order gradient scalar damage model. This model can be summarised by the injective relation between the stress and strain tensors, $\boldsymbol{\sigma}$ and $\boldsymbol{\epsilon}$, respectively:

$$\boldsymbol{\sigma} = (1 - \omega)\mathbf{D}^{\mathrm{e}} : \boldsymbol{\epsilon} \tag{50}$$

with ω a scalar damage variable which grows from zero to one (at complete loss of integrity) and $\mathbf{D}^{\mathrm{e}}$ the fourth-order elastic stiffness tensor. This total stress-strain relation is complemented by a damage loading function f, which reads

$$f = \tilde{\epsilon} - \kappa \tag{51}$$

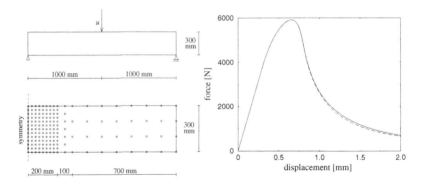

Figure 20. Left: Three-point bending beam and node distribution (for a symmetric half of the beam). Right: Load-displacement curves for three-point bending beam. Comparison between the second-order implicit gradient damage model (dashed line) and the fourth-order implicit gradient damage model (solid line), see Askes *et al.* (2000)

with the equivalent strain $\tilde{\epsilon}$ a scalar-valued function of the strain tensor, and κ a history variable. The damage loading function f and the rate of the history variable, $\dot{\kappa}$, have to satisfy the discrete Kuhn-Tucker loading-unloading conditions

$$f \leq 0, \quad \dot{\kappa} \geq 0, \quad \dot{\kappa} f = 0 \tag{52}$$

The history parameter κ starts at a damage threshold level κ_i and is updated by the requirement that during damage growth $f = 0$. Damage growth occurs according to an evolution law such that $\omega = \omega(\kappa)$, which can be determined from a uniaxial test.

In a non-local generalisation, the loading function (51) is replaced by

$$f = \bar{\epsilon} - \kappa \tag{53}$$

where the non-local equivalent strain $\bar{\epsilon}$ follows from the solution of the partial differential equation

$$\bar{\epsilon} - c_1 \nabla^2 \bar{\epsilon} - c_2 \nabla^4 \bar{\epsilon} = \tilde{\epsilon} \tag{54}$$

with c_1 and c_2 two gradient constants, with is assumed to hold on the entire domain. Evidently, even after order reduction by partial integration, $\mathcal{C}^1$-continuous shape functions are necessary for the interpolation of the non-local strain $\bar{\epsilon}$, with all the computational inconveniences that come to it when finite elements are employed. Here, meshfree methods offer a distinct advantage, since they can be easily constructed such that they incorporate $\mathcal{C}^\infty$-continuous shape functions. In Fig. 20 the element-free Galerkin method has been used to solve the damage evolution that is described by the fourth-order gradient scalar damage model of Eqs. (50), (52), (53) and (54) to predict the damage evolution in a three-point bending beam (Askes *et al.*, 2000). In both cases, a quadratic convergence behaviour was obtained when using a properly linearised tangent stiffness matrix. Interestingly, the differences between the fourth-order and the second-order ($c_2 = 0$) gradient damage models are almost negligible.

The size of the support of a node relative to the nodal spacing determines the properties of meshfree methods. When the support is made equal to the nodal spacing, shape functions as obtained in meshfree methods can become identical to finite element shape functions, thus showing that meshfree methods *encompass* finite element methods as a subclass. On the other hand, larger supports lead to shape functions in meshfree methods that are similar to higher-order polynomials, even if the base vector $\mathbf{p}(\mathbf{x})$ contains only constant and linear terms. Finally, it is noted that, although meshfree methods can straightforwardly accommodate cohesive-zone models, such computations seem absent in the literature.

7.   The partition-of-unity concept

Out of the research into meshless methods, a method has emerged in which a discontinuity in the displacement field is captured exactly. It has the added benefit that it can be used advantageously at different scales, from microscopic to macroscopic analyses. The method exploits the partition-of-unity property of finite element shape functions (Babuška and Melenk, 1997). A collection of functions $\phi_\imath$, associated with nodes $\imath$, form a partition of unity if $\sum_{\imath=1}^{n} \phi_\imath(\mathbf{x}) = 1$ with n the number of discrete nodal points. For a set of functions $\phi_\imath$ that satisfy this property, a field u can be interpolated as follows:

$$u(\mathbf{x}) = \sum_{\imath=1}^{n} \phi_\imath(\mathbf{x}) \left(\bar{a}_\imath + \sum_{\jmath=1}^{m} \psi_\jmath(\mathbf{x})\tilde{a}_{\imath\jmath} \right) \tag{55}$$

with $\bar{a}_\imath$ the 'regular' nodal degrees-of-freedom, $\psi_\jmath(\mathbf{x})$ the enhanced basis terms, and $\tilde{a}_{\imath\jmath}$ the additional degrees-of-freedom at node $\imath$ which represent the amplitudes of the $\jmath$th enhanced basis term $\psi_\jmath(\mathbf{x})$. In conventional finite element notation we can thus interpolate a displacement field as

$$\mathbf{u} = \mathbf{N}(\bar{\mathbf{a}} + \tilde{\mathbf{N}}\tilde{\mathbf{a}}) \tag{56}$$

where $\mathbf{N}$ contains the standard shape functions, $\tilde{\mathbf{N}}$ the enhanced basis terms and $\bar{\mathbf{a}}$ and $\tilde{\mathbf{a}}$ collect the conventional and the additional nodal degrees-of-freedom, respectively. A displacement field that contains a single discontinuity, Fig. 21, can be represented by choosing (Belytschko and Black, 1999; Moës *et al.*, 1999):

$$\tilde{\mathbf{N}} = \mathcal{H}_{\Gamma_d}\mathbf{I} \tag{57}$$

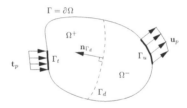

Figure 21.   Body composed of continuous displacement fields at each side of the discontinuity Γ_d

Substitution into Eq. (56) gives

$$\mathbf{u} = \underbrace{\mathbf{N}\bar{\mathbf{a}}}_{\bar{\mathbf{u}}} + \mathcal{H}_{\Gamma_d}\underbrace{\mathbf{N}\tilde{\mathbf{a}}}_{\tilde{\mathbf{u}}} \tag{58}$$

Identifying $\bar{\mathbf{u}} = \mathbf{N}\bar{\mathbf{a}}$ and $\tilde{\mathbf{u}} = \mathbf{N}\tilde{\mathbf{a}}$ we observe that Eq. (58) exactly describes a displacement field that is crossed by a single discontinuity, but is otherwise continuous. Accordingly, the partition-of-unity property of finite element shape functions can be used in a straightforward fashion to incorporate discontinuities in a manner that preserves their discontinuous character.

As before, we take the balance of momentum

$$\nabla \cdot \boldsymbol{\sigma} + \rho\mathbf{g} = \mathbf{0} \tag{59}$$

as point of departure and multiply this identity by test functions $\boldsymbol{w}$, taking them from the same space as the trial functions for $\mathbf{u}$,

$$\boldsymbol{w} = \bar{\boldsymbol{w}} + \mathcal{H}_{\Gamma_d}\tilde{\boldsymbol{w}} \tag{60}$$

Applying the divergence theorem and requiring that this identity holds for arbitrary $\bar{\boldsymbol{w}}$ and $\tilde{\boldsymbol{w}}$ yields the following set of coupled equations:

$$\int_{\Omega} \nabla^{\mathrm{sym}}\bar{\boldsymbol{w}} : \boldsymbol{\sigma}\mathrm{d}\Omega = \int_{\Omega} \bar{\boldsymbol{w}} \cdot \rho\mathbf{g}\mathrm{d}\Omega + \int_{\Gamma} \bar{\boldsymbol{w}} \cdot \mathbf{t}\mathrm{d}\Gamma \tag{61}$$

$$\int_{\Omega^+} \nabla^{\mathrm{sym}}\tilde{\boldsymbol{w}} : \boldsymbol{\sigma}\mathrm{d}\Omega + \int_{\Gamma_d} \tilde{\boldsymbol{w}} \cdot \mathbf{t}_i\mathrm{d}\Gamma = \int_{\Omega^+} \tilde{\boldsymbol{w}} \cdot \rho\mathbf{g}\mathrm{d}\Omega + \int_{\Gamma} \mathcal{H}_{\Gamma_d}\tilde{\boldsymbol{w}} \cdot \mathbf{t}\mathrm{d}\Gamma \tag{62}$$

where in the volume integrals the Heaviside function has been eliminated by a change of the integration domain from Ω to Ω^+. With the standard interpolation:

$$\begin{aligned} \bar{\mathbf{u}} &= \mathbf{N}\bar{\mathbf{a}}, \quad \tilde{\mathbf{u}} = \mathbf{N}\tilde{\mathbf{a}} \\ \bar{\boldsymbol{w}} &= \mathbf{N}\bar{\mathbf{w}}, \quad \tilde{\boldsymbol{w}} = \mathbf{N}\tilde{\mathbf{w}} \end{aligned} \tag{63}$$

and requiring that the resulting equations must hold for any admissible $\bar{\mathbf{w}}$ and $\tilde{\mathbf{w}}$, we obtain the discrete format:

$$\int_{\Omega} \mathbf{B}^T \boldsymbol{\sigma} d\Omega = \int_{\Omega} \rho \mathbf{B}^T \mathbf{g} d\Omega + \int_{\Gamma} \mathbf{N}^T \mathbf{t} d\Gamma \tag{64}$$

$$\int_{\Omega^+} \mathbf{B}^T \boldsymbol{\sigma} d\Omega + \int_{\Gamma_d} \mathbf{N}^T \mathbf{t}_d d\Gamma = \int_{\Omega^+} \rho \mathbf{B}^T \mathbf{g} d\Omega + \int_{\Gamma} \mathcal{H}_{\Gamma_d} \mathbf{N}^T \mathbf{t} d\Gamma \tag{65}$$

After linearisation, the following matrix-vector equation is obtained

$$\begin{bmatrix} \mathbf{K}_{\bar{a}\bar{a}} & \mathbf{K}_{\bar{a}\tilde{a}} \\ \mathbf{K}_{\tilde{a}\bar{a}} & \mathbf{K}_{\tilde{a}\tilde{a}} \end{bmatrix} \begin{pmatrix} d\bar{a} \\ d\tilde{a} \end{pmatrix} = \begin{pmatrix} \mathbf{f}_{\bar{a}}^{ext} - \mathbf{f}_{\bar{a}}^{int} \\ \mathbf{f}_{\tilde{a}}^{ext} - \mathbf{f}_{\tilde{a}}^{int} \end{pmatrix} \tag{66}$$

with $\mathbf{f}_{\bar{a}}^{int}, \mathbf{f}_{\tilde{a}}^{int}$ given by the left-hand sides of Eqs. (61)–(62), $\mathbf{f}_{\bar{a}}^{ext}, \mathbf{f}_{\tilde{a}}^{ext}$ given by the right-hand sides of Eqs. (61)–(62) and

$$\mathbf{K}_{\bar{a}\bar{a}} = \int_{\Omega} \mathbf{B}^T \mathbf{D} \mathbf{B} d\Omega \tag{67}$$

$$\mathbf{K}_{\bar{a}\tilde{a}} = \int_{\Omega^+} \mathbf{B}^T \mathbf{D} \mathbf{B} d\Omega \tag{68}$$

$$\mathbf{K}_{\tilde{a}\bar{a}} = \int_{\Omega^+} \mathbf{B}^T \mathbf{D} \mathbf{B} d\Omega \tag{69}$$

$$\mathbf{K}_{\tilde{a}\tilde{a}} = \int_{\Omega^+} \mathbf{B}^T \mathbf{D} \mathbf{B} d\Omega + \int_{\Gamma_d} \mathbf{N}^T \mathbf{T} d\Gamma \tag{70}$$

If the material tangential stiffness matrices of the bulk and the interface, $\mathbf{D}$ and $\mathbf{T}$ respectively, are symmetric, the total tangential stiffness matrix remains symmetric. It is emphasised that in this concept, the additional degrees-of-freedom *cannot* be condensed at element level, because it is node-oriented and not element-oriented. It is this property which makes it possible to represent a discontinuity such that it is continuous at interelement boundaries.

The partition-of-unity property of finite element shape functions is a powerful method to introduce cohesive surfaces in continuum finite elements (Wells and Sluys, 2001; Wells *et al.*, 2002; Moës and Belytschko, 2002). Using the interpolation of Eq. (58) the relative displacement at the discontinuity Γ_d is obtained as

$$\mathbf{v} = \tilde{\mathbf{u}} \mid_{\mathbf{x} \in \Gamma_d} \tag{71}$$

and the tractions at the discontinuity are derived from Eq. (25). A key feature of the method is the possibility of extending a (cohesive) crack during the calculation in an arbitrary direction, independent of the structure of the underlying finite element mesh. It is also interesting to note that the field $\tilde{u}$ does not have to be constant. The only requirement that is imposed is continuity.

When the discontinuity coincides with a side of the element, the traditional interface element formulation is retrieved. For this, we expand the term in $\mathbf{K}_{\tilde{a}\tilde{a}}$ which relates to the discontinuity as

$$\int_{\Gamma_i} \mathbf{H}^T \mathbf{T} \mathbf{H} d\Gamma = \begin{bmatrix} \mathbf{K}_n & 0 & 0 \\ 0 & \mathbf{K}_s & 0 \\ 0 & 0 & \mathbf{K}_t \end{bmatrix} \tag{72}$$

with $\mathbf{K}_\pi = d_\pi \mathbf{h}^T \mathbf{h}$ (cf. Simone, 2004), which closely resembles Eqs. (31)–(32). Defining the sum of the nodal displacements $\bar{a}$ and $\tilde{a}$ as primary variable a on the + side of the interface and setting $a = \bar{a}$ at the - side and rearranging then leads to the standard interface formulation.

However, even though formally the matrices can coincide for the partition–of–unity based method and for the conventional interface formulation, both concepts are quite different. Indeed, simulations of delamination using the partition-of-unity property of finite element shape functions offer advantages. Because the discontinuity does not have to be inserted a priori, no (dummy) stiffness is needed in the elastic regime. Indeed, there does not have to be an elastic regime, since the discontinuity can be activated at the onset of cracking. Consequently, the issue of spurious traction oscillations in the elastic phase becomes irrelevant. Also, the lines of the potential delamination planes no longer have to coincide with element boundaries. They can lie at arbitrary locations inside elements and unstructured meshes can be used.

The above approach for capturing discontinuities can be generalised to large displacement gradients in a straightforward and consistent manner. To this end, one extends Eq. (58) as

$$\mathbf{x} = \mathbf{X} + \bar{u} + \mathcal{H}_{\Gamma_d^0} \tilde{u} \tag{73}$$

with $\mathcal{H}_{\Gamma_d^0}$ the Heaviside function at the interface in the reference configuration, Γ_d^0. The deformation gradient follows by differentiation:

$$\mathbf{F} = \bar{\mathbf{F}} + \mathcal{H}_{\Gamma_d^0} \tilde{\mathbf{F}} + \delta_{\Gamma_d^0}(\tilde{u} \otimes \mathbf{n}_{\Gamma_d^0}) \tag{74}$$

with $\bar{\mathbf{F}} = \boldsymbol{i} + \partial\bar{u}/\partial\mathbf{X}$, $\tilde{\mathbf{F}} = \partial\tilde{u}/\partial\mathbf{X}$ and $\delta_{\Gamma_d^0}$ the Dirac function at the interface in the reference configuration.

With aid of Nanson's relation for the normal $\mathbf{n}$ to a surface Γ:

$$\mathbf{n} = \det \mathbf{F}(\mathbf{F}^{-T})\mathbf{n}_{\Gamma^0} \frac{d\Gamma^0}{d\Gamma} \tag{75}$$

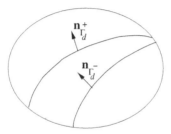

Figure 22. Body crossed by a discontinuity Γ_d with normals $\mathbf{n}_{\Gamma_d^-}$ and $\mathbf{n}_{\Gamma_d^+}$ at both sides of the discontinuity

the expressions for the normals at the - side and at the + side of the interface can be derived

$$\mathbf{n}_{\Gamma_d^-} = \det \bar{\mathbf{F}}(\bar{\mathbf{F}}^{-T})\mathbf{n}_{\Gamma_d^0}\frac{\mathrm{d}\Gamma_d^0}{\mathrm{d}\Gamma_d^-} \tag{76a}$$

$$\mathbf{n}_{\Gamma_d^+} = \det(\bar{\mathbf{F}} + \tilde{\mathbf{F}})(\bar{\mathbf{F}} + \tilde{\mathbf{F}})^{-T}\mathbf{n}_{\Gamma_d^0}\frac{\mathrm{d}\Gamma_d^0}{\mathrm{d}\Gamma_d^+} \tag{76b}$$

see Fig. 22. Distinction between $\mathbf{n}_{\Gamma_d^-}$ and $\mathbf{n}_{\Gamma_d^+}$ is possible because $\tilde{\mathbf{u}}$ is not spatially constant. In the cohesive-zone approach, interface tractions $\mathbf{t}_i$ are transmitted between Γ^- and Γ^+ with different normals $\mathbf{n}_{\Gamma_d^-}$ and $\mathbf{n}_{\Gamma_d^+}$. In a heuristic assumption, it has been assumed that an average normal can be defined for use within the cohesive-zone model (Wells *et al.*, 2002):

$$\mathbf{n}_{\Gamma_d^*} = \det(\bar{\mathbf{F}} + \frac{1}{2}\tilde{\mathbf{F}})(\bar{\mathbf{F}} + \frac{1}{2}\tilde{\mathbf{F}})^{-T}\mathbf{n}_{\Gamma_d^0}\frac{\mathrm{d}\Gamma_d^0}{\mathrm{d}\Gamma_d^*} \tag{77}$$

We now recall the equilibrium equation in the current configuration, cf. Eq. (9):

$$\nabla_{\mathbf{x}}\cdot\boldsymbol{\sigma} + \rho\mathbf{g} = \mathbf{0}$$

In a Bubnov-Galerkin method the test functions $\boldsymbol{w}$ for a single discontinuity are given by

$$\boldsymbol{w} = \bar{\boldsymbol{w}} + \mathcal{H}_{\Gamma_d^0}\tilde{\boldsymbol{w}} \tag{78}$$

Multiplying with this test function, integrating over the current domain Ω and requiring that the result holds for arbitrary $\bar{\boldsymbol{w}}$ and $\tilde{\boldsymbol{w}}$ yields

$$\int_{\Omega}\nabla_{\mathbf{x}}\bar{\boldsymbol{w}} : \boldsymbol{\sigma}\mathrm{d}\Omega = \int_{\Omega}\bar{\boldsymbol{w}}\cdot\rho\mathbf{g}\mathrm{d}\Omega + \int_{\Gamma}\bar{\boldsymbol{w}}\cdot\mathbf{t}\mathrm{d}\Gamma \tag{79a}$$

$$\int_{\Omega^+}\nabla_{\mathbf{x}}\tilde{\boldsymbol{w}} : \boldsymbol{\sigma}\mathrm{d}\Omega + \int_{\Gamma_d}\tilde{\boldsymbol{w}}\cdot\mathbf{t}_i\mathrm{d}\Gamma = \int_{\Omega^+}\tilde{\boldsymbol{w}}\cdot\rho\mathbf{g}\mathrm{d}\Omega + \int_{\Gamma}\mathcal{H}_{\Gamma_d}\tilde{\boldsymbol{w}}\cdot\mathbf{t}\mathrm{d}\Gamma \tag{79b}$$

with the subscript x signifying differentiation with respect to the current configuration and $t_d = n_{\Gamma_d^*} \cdot \boldsymbol{\sigma}$ the traction at the discontinuity in the current configuration. With a standard interpolation:

$$\bar{w} = \mathbf{N}\bar{\mathbf{w}}, \qquad \tilde{w} = \mathbf{N}\tilde{\mathbf{w}} \tag{80}$$

where $\mathbf{N}$ contains the interpolation polynomials and $\bar{\mathbf{w}}$ and $\tilde{\mathbf{w}}$ contain the discrete values for the test functions, the discrete format of Eqs. (79a)–(79b) reads

$$\int_{\Omega} \mathbf{B}^{\mathrm{T}} \boldsymbol{\sigma} \mathrm{d}\Omega = \int_{\Omega} \rho \mathbf{B}^{\mathrm{T}} \mathbf{g} \mathrm{d}\Omega + \int_{\Gamma} \mathbf{N}^{\mathrm{T}} \mathbf{t} \mathrm{d}\Gamma \tag{81a}$$

$$\int_{\Omega^+} \mathbf{B}^{\mathrm{T}} \boldsymbol{\sigma} \mathrm{d}\Omega + \int_{\Gamma_d} \mathbf{N}^{\mathrm{T}} \mathbf{t}_i \mathrm{d}\Gamma = \int_{\Omega^+} \rho \mathbf{B}^{\mathrm{T}} \mathbf{g} \mathrm{d}\Omega + \int_{\Gamma} \mathcal{H}_{\Gamma_d} \mathbf{N}^{\mathrm{T}} \mathbf{t} \mathrm{d}\Gamma \tag{81b}$$

After substitution of the constitutive relations for the plies and that for the interface, and transforming back to the reference configuration, a nonlinear set of algebraic equations results, which can be solved in a standard manner using an incremental-iterative procedure. If a Newton–Raphson procedure is used, these equations have to be linearised in order to derive the structural tangential stiffness matrix (Wells *et al.*, 2002).

To exemplify the possibilities of this approach to model the combined failure mode of delamination growth and local buckling we consider the double cantilever beam of Fig. 23 with an initial delamination length $a = 10\ mm$. Both layers are made of the same material with Young's modulus $E = 1,35,000\ N/mm^2$ and Poisson's ratio $\nu = 0.18$. Due to symmetry in the geometry of the model and the applied loading, delamination propagation can be modelled with an exponential mode-I decohesion law:

$$t_{\mathrm{dis}}^n = t_{\mathrm{ult}} \exp\left(-\frac{t_{\mathrm{ult}}}{\mathcal{G}_c} v_{\mathrm{dis}}^n\right) \tag{82}$$

where t_{dis}^n and v_{dis}^n are the normal traction and displacement jump respectively. The ultimate traction t_{ult} is equal to $50\ N/mm^2$, the work of separation is $\mathcal{G}_c = 0.8\ N/mm$.

This case, in which failure is a consequence of a combination of delamination growth and structural instability, has been analysed using conventional interface elements by Allix and Corigliano (1999). The beam is subjected to an axial compressive force $2P$, while two small perturbing forces P_0 are applied to trigger the buckling mode. Two finite element discretisations have been employed, a fine mesh with three elements over the thickness and 250 elements along the length of the beam, and a coarse mesh with only one (!) element over the thickness and 100 elements along the length. Figure 24 shows that the calculation with the coarse mesh approaches the results for the fine mesh closely. For instance, the numerically calculated buckling load is in good agreement

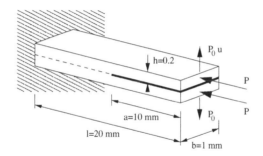

Figure 23. Double cantilever beam with initial delamination under compression

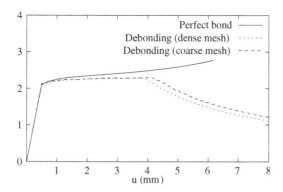

Figure 24. Load-displacement curves for delamination-buckling test

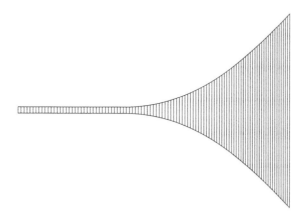

Figure 25. Deformation of coarse mesh after buckling and delamination growth (true scale)

with the analytical solution. Steady-state delamination growth starts around a lateral displacement $u = 4\ mm$. From this point onwards, delamination growth interacts with geometrical instability. Figure 25 presents the deformed beam for the coarse mesh at a tip displacement $u = 6\ mm$. Note that the displacements are plotted at true scale, but that the difference in displacement between the upper and lower parts of the beam is for the major part due to the delamination and that the strains remain small.

8.   Delamination in a solid-like shell element

The excellent results obtained in this example for the coarse discretisation have motivated the development of a layered plate/shell element in which delaminations can occur inside the element between each of the layers (Remmers *et al.*, 2003). Because of the absence of rotational degrees-of-freedom, the solid-like shell element was taken as a point of departure. The shell of Fig. 26 is crossed by a discontinuity surface Γ_d^0 which is assumed to be parallel to the mid-surface of the thick shell. The displacement field $\phi(\xi, \eta, \zeta)$ can now be regarded as a continuous regular field $\bar{\phi}$ with an additional continuous field $\tilde{\phi}$ that determines the magnitude of the displacement jump. The position of a material point in the deformed configuration can then be written as

$$\mathbf{x} = \mathbf{X} + \bar{\phi} + \mathcal{H}_{\Gamma_d^0}\tilde{\phi} \tag{83}$$

Since the displacement field ϕ is a function of the variables $\mathbf{u}_t$, $\mathbf{u}_b$ and w, we need to decompose these three terms as

$$\mathbf{u}_t = \bar{\mathbf{u}}_t + \mathcal{H}_{\Gamma_d^0}\tilde{\mathbf{u}}_t$$
$$\mathbf{u}_b = \bar{\mathbf{u}}_b + \mathcal{H}_{\Gamma_d^0}\tilde{\mathbf{u}}_b \tag{84}$$
$$w = \bar{w} + \mathcal{H}_{\Gamma_d^0}\tilde{w}$$

Inserting Eq. (84) into Eqs. (38)–(40) gives

$$\mathbf{u}_0 = \bar{\mathbf{u}}_0 + \mathcal{H}_{\Gamma_d^0}\tilde{\mathbf{u}}_0$$
$$\mathbf{u}_1 = \bar{\mathbf{u}}_1 + \mathcal{H}_{\Gamma_d^0}\tilde{\mathbf{u}}_1 \tag{85}$$
$$\mathbf{u}_2 = \bar{\mathbf{u}}_2 + \mathcal{H}_{\Gamma_d^0}\tilde{\mathbf{u}}_2$$

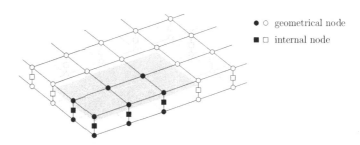

● ○ geometrical node

■ □ internal node

Figure 26.   Enhanced nodes (black) whose support contains a discontinuity (grey surface). The nodes on the edge of the discontinuity are not enhanced in order to ensure a zero delamination opening at the tip

where

$$\bar{\mathbf{u}}_0 = \frac{1}{2}\left[\bar{\mathbf{u}}_t + \bar{\mathbf{u}}_b\right] \qquad \tilde{\mathbf{u}}_0 = \frac{1}{2}\left[\tilde{\mathbf{u}}_t + \tilde{\mathbf{u}}_b\right]$$

$$\bar{\mathbf{u}}_1 = \frac{1}{2}\left[\bar{\mathbf{u}}_t - \bar{\mathbf{u}}_b\right] \qquad \tilde{\mathbf{u}}_1 = \frac{1}{2}\left[\tilde{\mathbf{u}}_t - \tilde{\mathbf{u}}_b\right] \qquad (86)$$

$$\bar{\mathbf{u}}_2 = \bar{w}\left[\mathbf{D} + \bar{\mathbf{u}}_1\right] \qquad \tilde{\mathbf{u}}_2 = \tilde{w}\left[\mathbf{D} + \bar{\mathbf{u}}_1 + \tilde{\mathbf{u}}_1\right] + \bar{w}\tilde{\mathbf{u}}_1$$

It is noted that the enhanced part of the internal stretch parameter $\mathbf{u}_2$, i.e. $\tilde{\mathbf{u}}_2$, contains regular variables as well as additional variables. The elaboration of the strains, the equilibrium equations and the linearisation follows standard lines (Remmers *et al.*, 2003).

The magnitude of the displacement jump at the discontinuity is governed by an additional set of degrees-of-freedom which are added to the existing nodes of the model. Figure 26 shows the activation of these additional sets of degrees of freedom for a given (static) delamination surface in the model. Both the geometrical and the internal nodes are enhanced when the corresponding element is crossed by the discontinuity. This implies that each geometrical node now contains three additional degrees-of-freedom next to the three regular ones, giving six degrees-of-freedom in total. Each internal node has one extra degree-of-freedom added to the single regular degree-of-freedom. As in the continuum elements, the discontinuity is assumed to always stretch through an entire element. This avoids the need for complicated algorithms to describe the stress state in the vicinity of a delamination front within an element. As a consequence, the discontinuity 'touches' the boundary of an element. The geometrical and internal nodes that support this boundary are not enhanced in order to assure a zero crack tip condition.

The example of Fig. 23 has been reanalysed with a mesh composed of eight node enhanced solid-like shell elements (Remmers *et al.*, 2003). Again, only one element in thickness direction has been used. In order to capture delamination growth correctly, the mesh has been refined locally. Figure 27 shows the lateral displacement u of the beam as a function of the external force P. The

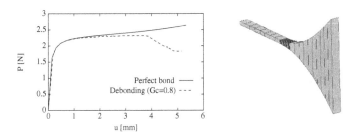

Figure 27. Load-displacement curve and deformations of shell model after buckling and delamination growth (true scale), (Remmers *et al.*, 2003)

load-displacement response for a specimen with a perfect bond (no delamination growth) is given as a reference. The numerically calculated buckling load is in agreement with the analytical solution. Steady delamination growth starts around a lateral displacement $u \approx 4\,mm$, which is in agreement with the previous simulations, e.g., by Allix and Corigliano (1999).

9.   Discontinuous Galerkin methods

Discontinuous Galerkin methods have classically been employed for the computation of fluid flow, e.g. Cockburn (2004). More recently, attention has been given to their potential use in solid mechanics, and especially for problems involving cracks (Mergheim *et al.*, 2004), or for constitutive models that incorporate spatial gradients (Wells *et al.*, 2004), such as gradient plasticity or gradient damage, e.g., (de Borst and Gutiérrez, 1999; Askes *et al.*, 2000). In the latter case, the fact that discontinous Galerkin methods do not require interelement continuity a priori, by-passes the requirement of $\mathcal{C}^1$-continuity on the damage or plastic multiplier field which plague the implementation of many gradient models in a continuous Galerkin finite element method. In the former case, use of a discontinuous Galerkin formalism can be an alternative way to avoid traction oscillations in the pre-cracking phase (Mergheim *et al.*, 2004).

For a discussion on the application of spatially discontinuous Galerkin to fracture it suffices by dividing the domain in two subdomains, Ω^- and Ω^+, separated by an interface Γ_i. In a standard manner, the balance of momentum (9) is multiplied by a test function w, and after application of the divergence theorem, we obtain

$$\int_{\Omega/\Gamma_i} \nabla^{\text{sym}} w : \sigma \mathrm{d}\Omega - \int_{\Gamma_i} w^+ \cdot t_i^+ \mathrm{d}\Gamma - \int_{\Gamma_i} w^- \cdot t_i^- \mathrm{d}\Gamma = \int_{\Omega/\Gamma_i} w \cdot \rho g \mathrm{d}\Omega + \int_{\Gamma} w \cdot t \mathrm{d}\Gamma$$

$$(87)$$

where the surface (line) integral on the external boundary Γ has been explicitly separated from that on the interface Γ_i. Prior to crack initiation, continuity of displacements and tractions must be enforced along Γ_i, at least in an approximate sense:

$$u^+ - u^- = 0 \qquad (88a)$$

$$t_i^+ + t_i^- = 0 \qquad (88b)$$

with $t_i^+ = n_{\Gamma_i}^+ \cdot \sigma^+$ and $t_i^- = n_{\Gamma_i}^- \cdot \sigma^-$. Assuming small displacement gradients, we can set $n_{\Gamma_i} = n_{\Gamma_i}^+ = -n_{\Gamma_i}^-$, so that the expressions for the interface tractions reduce to $t_i^+ = n_{\Gamma_i} \cdot \sigma^+$ and $t_i^- = -n_{\Gamma_i} \cdot \sigma^-$.

A classical procedure to enforce conditions (88a)–(88b) is to use Lagrange multipliers. Then,

$$\lambda = t_i^+ = -t_i^- \qquad (89)$$

along Γ_i, and Eq. (87) transforms into:

$$\int_{\Omega/\Gamma_i} \nabla^{\text{sym}}\boldsymbol{w} : \boldsymbol{\sigma}\mathrm{d}\Omega - \int_{\Gamma_i} (\boldsymbol{w}^+ - \boldsymbol{w}^-) \cdot \boldsymbol{\lambda}\mathrm{d}\Gamma = \int_{\Omega/\Gamma_i} \boldsymbol{w} \cdot \rho\mathbf{g}\mathrm{d}\Omega + \int_{\Gamma} \boldsymbol{w} \cdot \mathbf{t}\mathrm{d}\Gamma \quad (90)$$

augmented with

$$\int_{\Gamma_i} \boldsymbol{z} \cdot (\mathbf{u}^+ - \mathbf{u}^-)\mathrm{d}\Gamma = 0 \quad (91)$$

$\boldsymbol{z}$ being the test function for the Lagrange multiplier field $\boldsymbol{\lambda}$. After discretisation, Eqs. (90) and (91) result in a set of algebraic equations that are of a stan-
dard mixed format and therefore give rise to difficulties when using solvers
without pivoting. For this reason, alternative expressions are often sought, in
which $\boldsymbol{\lambda}$ is directly expressed in terms of the interface tractions $\mathbf{t}_i^-$ and $\mathbf{t}_i^+$. One
such possibility is to enforce Eq. (88b) pointwise, so that Eq. (89) is replaced
by:

$$\boldsymbol{\lambda} = -\mathbf{t}_i \quad (92)$$

and one obtains

$$\int_{\Omega/\Gamma_i} \nabla^{\text{sym}}\boldsymbol{w} : \boldsymbol{\sigma}\mathrm{d}\Omega + \int_{\Gamma_i} (\boldsymbol{w}^+ - \boldsymbol{w}^-) \cdot \mathbf{t}_i\mathrm{d}\Gamma = \int_{\Omega/\Gamma_i} \boldsymbol{w} \cdot \rho\mathbf{g}\mathrm{d}\Omega + \int_{\Gamma} \boldsymbol{w} \cdot \mathbf{t}\mathrm{d}\Gamma \quad (93)$$

With aid of relation (23) between the relative displacements $\mathbf{v} = \mathbf{u}^+ - \mathbf{u}^-$
and the nodal displacements at both sides of the interface Γ_i, and the interface
traction-relative displacement relation (26), the second term on the left-hand
side can be elaborated in a discrete format as

$$\int_{\Gamma_i} (\boldsymbol{w}^+ - \boldsymbol{w}^-) \cdot \mathbf{t}_i\mathrm{d}\Gamma = \mathbf{w}^{\text{T}} \left(\int_{\Gamma_i} \mathbf{B}_i^{\text{T}}\mathbf{T}\mathbf{B}_i\mathrm{d}\Gamma \right) \mathbf{a} \quad (94)$$

which, not surprisingly, has exactly the same format as obtained for a conventional interface element.

Another possibility for the replacement of $\boldsymbol{\lambda}$ by an explicit function of the
tractions is to take the average of the stresses at both sides of the interface:

$$\boldsymbol{\lambda} = \frac{1}{2} \mathbf{n}_{\Gamma_i} \cdot (\boldsymbol{\sigma}^+ + \boldsymbol{\sigma}^-) \quad (95)$$

The surface integrals for the interface in Eq. (90) can now be reworked as:

$$\int_{\Gamma_i} (\boldsymbol{w}^+ - \boldsymbol{w}^-) \cdot \boldsymbol{\lambda}\mathrm{d}\Gamma = \int_{\Gamma_i} \frac{1}{2}(\boldsymbol{w}^+ - \boldsymbol{w}^-) \cdot \mathbf{n}_{\Gamma_i} \cdot (\boldsymbol{\sigma}^+ + \boldsymbol{\sigma}^-)\mathrm{d}\Gamma \quad (96)$$

To ensure a proper conditioning of the discretised equations, one has to add Eq. (91), so that the modified form of Eq. (87) finally becomes

$$\int_{\Omega/\Gamma_i} \nabla^{\text{sym}} \boldsymbol{w} : \boldsymbol{\sigma} \mathrm{d}\Omega - \int_{\Gamma_i} \frac{1}{2}(\boldsymbol{w}^+ - \boldsymbol{w}^-) \cdot \mathbf{n}_{\Gamma_i} \cdot (\boldsymbol{\sigma}^+ + \boldsymbol{\sigma}^-)\mathrm{d}\Gamma -$$

$$\alpha \int_{\Gamma_i} \frac{1}{2}(\nabla^{\text{sym}} \boldsymbol{w}^+ + \nabla^{\text{sym}} \boldsymbol{w}^-) : \mathbf{D} \cdot \mathbf{n}_{\Gamma_i} \cdot (\mathbf{u}^+ - \mathbf{u}^-)\mathrm{d}\Gamma = \qquad (97)$$

$$\int_{\Omega/\Gamma_i} \boldsymbol{w} \cdot \rho \mathbf{g} \mathrm{d}\Omega + \int_{\Gamma} \boldsymbol{w} \cdot \mathbf{t} \mathrm{d}\Gamma$$

To ensure symmetry, $\alpha = 1$, but then a diffusionlike term, $\int_{\Gamma_i} \tau (\boldsymbol{w}^+ - \boldsymbol{w}^-) \cdot (\mathbf{u}^+ - \mathbf{u}^-)\mathrm{d}\Gamma$ has to be added to ensure numerical stability (Nitsche, 1970). The numerical parameter $\tau = \mathcal{O}(|k|/h)$, with $|k|$ a suitable norm of the diffusionlike matrix that results from elaborating this term and h a measure of the grid density. For the unsymmetric choice $\alpha = -1$, addition of a diffusionlike term may not be necessary (Baumann and Oden, 1999).

With a standard interpolation on both Ω^- and Ω^+ and requiring that the resulting equations hold for any admissible $\boldsymbol{w}$, we obtain the discrete format:

$$\int_{\Omega^-} \mathbf{B}^{\text{T}} \boldsymbol{\sigma} \mathrm{d}\Omega + \int_{\Gamma_i} \frac{1}{2}\mathbf{N}^{\text{T}} \mathbf{n}_{\Gamma_i}^{\text{T}} (\boldsymbol{\sigma}^+ + \boldsymbol{\sigma}^-)\mathrm{d}\Gamma -$$

$$\alpha \int_{\Gamma_i} \frac{1}{2}\mathbf{B}^{\text{T}} \mathbf{D} \mathbf{n}_{\Gamma_i}^{\text{T}} (\mathbf{u}^+ - \mathbf{u}^-)\mathrm{d}\Gamma = \int_{\Omega^-} \mathbf{B}^{\text{T}} \rho \mathbf{g} \mathrm{d}\Omega + \int_{\Gamma} \mathbf{N}^{\text{T}} \mathbf{t} \mathrm{d}\Gamma$$

$$\int_{\Omega^+} \mathbf{B}^{\text{T}} \boldsymbol{\sigma} \mathrm{d}\Omega - \int_{\Gamma_i} \frac{1}{2}\mathbf{N}^{\text{T}} \mathbf{n}_{\Gamma_i}^{\text{T}} (\boldsymbol{\sigma}^+ + \boldsymbol{\sigma}^-)\mathrm{d}\Gamma - \qquad (98)$$

$$\alpha \int_{\Gamma_i} \frac{1}{2}\mathbf{B}^{\text{T}} \mathbf{D} \mathbf{n}_{\Gamma_i}^{\text{T}} (\mathbf{u}^+ - \mathbf{u}^-)\mathrm{d}\Gamma = \int_{\Omega^+} \mathbf{B}^{\text{T}} \rho \mathbf{g} \mathrm{d}\Omega + \int_{\Gamma} \mathbf{N}^{\text{T}} \mathbf{t} \mathrm{d}\Gamma$$

with $\mathbf{n}_{\Gamma_i}$ now written in a matrix form:

$$\mathbf{n}_{\Gamma_i}^{\text{T}} = \begin{bmatrix} n_x & 0 & 0 & n_y & 0 & n_z \\ 0 & n_y & 0 & n_x & n_y & 0 \\ 0 & 0 & n_z & 0 & n_z & n_x \end{bmatrix} \qquad (99)$$

where n_x, n_y, n_z are the components of the vector $\mathbf{n}_{\Gamma_i}$. After linearisation, one obtains

$$\begin{bmatrix} \mathbf{K}^{--} & \mathbf{K}^{-+} \\ \mathbf{K}^{+-} & \mathbf{K}^{++} \end{bmatrix} \begin{pmatrix} \mathrm{d}\mathbf{a}^- \\ \mathrm{d}\mathbf{a}^+ \end{pmatrix} = \begin{pmatrix} \mathbf{f}_{\mathbf{a}^-}^{ext} - \mathbf{f}_{\mathbf{a}^-}^{int} \\ \mathbf{f}_{\mathbf{a}^+}^{ext} - \mathbf{f}_{\mathbf{a}^+}^{int} \end{pmatrix} \qquad (100)$$

with $\mathbf{a}^+, \mathbf{a}^-$ arrays that contain the nodal values of the displacements at the minus and the plus side of the interface, respectively, with $\mathbf{f}_{\mathbf{a}^-}^{ext}$, $\mathbf{f}_{\mathbf{a}^+}^{ext}$ and $\mathbf{f}_{\mathbf{a}^-}^{int}$,

$\mathbf{f}_{a^+}^{int}$ the right and left-hand sides of Eq. (98), respectively, and the submatrices defined by

$$\mathbf{K}^{--} = \int_{\Omega^-} \mathbf{B}^{\mathrm{T}} \mathbf{D} \mathbf{B} d\Omega + \frac{1}{2} \int_{\Gamma_i} \mathbf{N}^{\mathrm{T}} \mathbf{n}_{\Gamma_i}^{\mathrm{T}} \mathbf{D} \mathbf{B} d\Gamma + \frac{1}{2} \alpha \int_{\Gamma_i} \mathbf{B}^{\mathrm{T}} \mathbf{D} \mathbf{n}_{\Gamma_i} \mathbf{N} d\Gamma$$

(101a)

$$\mathbf{K}^{-+} = \frac{1}{2} \int_{\Gamma_i} \mathbf{N}^{\mathrm{T}} \mathbf{n}_{\Gamma_i}^{\mathrm{T}} \mathbf{D} \mathbf{B} d\Gamma - \frac{1}{2} \alpha \int_{\Gamma_i} \mathbf{B}^{\mathrm{T}} \mathbf{D} \mathbf{n}_{\Gamma_i} \mathbf{N} d\Gamma \qquad (101b)$$

$$\mathbf{K}^{+-} = \frac{1}{2} \alpha \int_{\Gamma_i} \mathbf{B}^{\mathrm{T}} \mathbf{D} \mathbf{n}_{\Gamma_i} \mathbf{N} d\Gamma - \frac{1}{2} \int_{\Gamma_i} \mathbf{N}^{\mathrm{T}} \mathbf{n}_{\Gamma_i}^{\mathrm{T}} \mathbf{D} \mathbf{B} d\Gamma \qquad (101c)$$

$$\mathbf{K}^{++} = \int_{\Omega^+} \mathbf{B}^{\mathrm{T}} \mathbf{D} \mathbf{B} d\Omega - \frac{1}{2} \int_{\Gamma_i} \mathbf{N}^{\mathrm{T}} \mathbf{n}_{\Gamma_i}^{\mathrm{T}} \mathbf{D} \mathbf{B} d\Gamma - \frac{1}{2} \alpha \int_{\Gamma_i} \mathbf{B}^{\mathrm{T}} \mathbf{D} \mathbf{n}_{\Gamma_i} \mathbf{N} d\Gamma$$

(101d)

References

Alfano, G. and Crisfield, M.A. (2001). Finite element interface models for the delamination analysis of laminated composites: mechanical and computational issues. *International Journal for Numerical Methods in Engineering*, **50**: 1701–1736.

Allix, O. and Ladevèze, P. (1992). Interlaminar interface modelling for the prediction of delamination. *Composite Structures*, **22**: 235–242.

Allix, O. and Corigliano, A. (1999). Geometrical and interfacial non-linearities in the analysis of delamination in composites. *International Journal of Solids and Structures*, **36**: 2189–2216.

Askes, H., Pamin, J. and de Borst, R. (2000). Dispersion analysis and element–free Galerkin solutions of second and fourth–order gradient–enhanced damage models. *International Journal for Numerical Methods in Engineering*, **49**: 811–832.

Babuška, I. and Melenk, J.M. (1997). The partition of unity method. *International Journal for Numerical Methods in Engineering*, **40**: 727–758.

Barenblatt, G.I. (1962). The mathematical theory of equilibrium cracks in brittle fracture. *Advances in Applied Mechanics*, **7**: 55–129.

Baumann, C.E. and Oden, J.T. (1999). A discontinuous *hp* finite element method for the Euler and Navier–Stokes problems. *International Journal for Numerical Methods in Fluids*, **31**: 79–95.

Belytschko, T. and Black, T. (1999). Elastic crack growth in finite elements with minimal remeshing. *International Journal for Numerical Methods in Engineering*, **45**: 601–620.

Belytschko, T., Lu, Y.Y. and Gu, L. (1994). Element–free Galerkin methods. *International Journal for Numerical Methods in Engineering*, **37**: 229–256.

Cockburn, B. (2004). Discontinuous Galerkin methods for computational fluid dynamics. In E. Stein, R. de Borst and T.J.R. Hughes (editors). *The Encyclopedia of Computational Mechanics*, Volume III, Chapter 4. Wiley, Chichester.

Corigliano, A. (1993). Formulation, identification and use of interface models in the numerical analysis of composite delamination. *International Journal of Solids and Structures*, **30**: 2779–2811.

de Borst, R. (2004). Damage, material instabilities, and failure. In E. Stein, R. de Borst, and T.J.R. Hughes (editors). *Encyclopedia of Computational Mechanics*, Volume 2, Chapter 10. Wiley, Chichester.

de Borst, R. and Gutiérrez, M.A. (1999). A unified framework for concrete damage and fracture models with a view to size effects. *International Journal of Fracture*, **95**: 261–277.

Dugdale, D.S. (1960). Yielding of steel sheets containing slits. *Journal of the Mechanics and Physics of Solids*, **8**: 100–108.

Feyel, F. and Chaboche, J.L. (2000). FE^2 multiscale approach for modelling the elastoviscoplastic behaviour of long fibre SiC/Ti composite materials. *Computer Methods in Applied Mechanics and Engineering*, **183**: 309–330.

Fleming, M., Chu, Y.A., Moran, B. and Belytschko, T. (1997). Enriched element-free Galerkin methods for crack tip fields. *International Journal for Numerical Methods in Engineering*, **40**: 1483–1504.

Hashagen, F. and de Borst, R. (2000). Numerical assessment of delamination in fibre metal laminates. *Computer Methods in Applied Mechanics and Engineering*, **185**: 141–159.

Krysl, P. and Belytschko, T. (1999). The element-free Galerkin method for dynamic propagation of arbitrary 3-D cracks. *International Journal for Numerical Methods in Engineering*, **44**: 767–800.

Ladevèze, P. and Lubineau, G. (2002). An enhanced mesomodel for laminates based on micromechanics. *Composites Science and Technology*, **62**: 533–541.

Mergheim, J., Kuhl, E. and P. Steinmann, P. (2004). A hybrid discontinuous Galerkin/interface method for the computational modelling of failure. *Communications in Numerical Methods in Engineering*, **20**: 511–519.

Moës, N. and Belytschko, T. (2002). Extended finite element method for cohesive crack growth. *Engineering Fracture Mechanics*, **69**: 813–833.

Moës, N., Dolbow, J. and Belytschko, T. (1999). A finite element method for crack growth without remeshing. *International Journal for Numerical Methods in Engineering*, **46**: 131–150.

Nayroles, B., Touzot, G. and Villon, P. (1992). Generalizing the finite element method: diffuse approximations and diffuse elements. *Computational Mechanics*, **10**: 307–318.

Nitsche, J.A. (1970). Über ein Variationsprinzip zur Lösung Dirichlet–Problemen bei Verwendung von Teilräumen, die keinen Randbedingungen unterworfen sind. *Abhandlungen des Mathematischen Seminars Universität Hamburg*, **36**: 9–15.

Parisch, H. (1995). A continuum-based shell theory for non-linear applications. *International Journal for Numerical Methods in Engineering*, **38**: 1855–1883.

Remmers, J.J.C. and de Borst, R. (2002). Delamination buckling of Fibre-Metal Laminates under compressive and shear loadings. *43rd AIAA/ASME/ASCE/AHS/ASC Structures, Structural Dynamics and Materials Conference*, Denver, Colorado, CD-ROM.

Remmers, J.J.C., Wells, G.N. and de Borst, R. (2003). A solid–like shell element allowing for arbitrary delaminations. *International Journal for Numerical Methods in Engineering*, **58**: 2013–2040.

Schellekens, J.C.J. and de Borst, R. (1992). On the numerical integration of interface elements. *International Journal for Numerical Methods in Engineering*, **36:** 43–66.

Schellekens, J.C.J. and de Borst, R. (1993). A non-linear finite element approach for the analysis of mode-I free edge delamination in composites. *International Journal of Solids and Structures*, **30:** 1239–1253.

Schellekens, J.C.J. and de Borst, R. (1994a). Free edge delamination in carbon-epoxy laminates: a novel numerical/experimental approach. *Composite Structures*, **28:** 357–373.

Schellekens, J.C.J. and de Borst, R. (1994b). The application of interface elements and enriched or rate-dependent continuum models to micro-mechanical analyses of fracture in composites. *Computational Mechanics*, **14:** 68–83.

Schipperen, J.H.A. and Lingen, F.J. (1999). Validation of two–dimensional calculations of free edge delamination in laminated composites. *Composite Structures*, **45:** 233–240.

Shivakumar, K. and Whitcomb, J. (1985). Buckling of a sublaminate in a quasi-isotropic composite laminate. *Journal of Composite Materials*, **19:** 2–18.

Simone, A. (2004). Partition of unity–based discontinuous elements for interface phenomena: computational issues. *Communications in Numerical Methods in Engineering*, **20:** 465–478.

Wells, G.N. and Sluys, L.J. (2001). A new method for modeling cohesive cracks using finite elements. *International Journal for Numerical Methods in Engineering*, **50:** 2667–2682.

Wells, G.N., de Borst, R. and Sluys, L.J. (2002). A consistent geometrically non–linear approach for delamination. *International Journal for Numerical Methods in Engineering*, **54:** 1333–1355.

Wells, G.N., Garikipati, K. and Molari, L. (2004). A discontinuous Galerkin formulation for a strain gradient–dependent damage model. *Computer Methods in Applied Mechanics and Engineering*, **193:** 3633–3645.

MICROMECHANICS OF COMPOSITES
Overall elastic properties

Ryszard Pyrz
Department of Mechanical Engineering
Aalborg University
Pontoppidanstræde 101
9220 Aalborg East
Denmark
rp@ime.aau.dk

Abstract   Composite materials are inherently inhomogeneous in terms of both elastic and inelastic properties. One consequence of this is that, on applying a load, a non-uniform distribution of stress is set up within the composite. Much effort has been devoted to understanding and predicting this distribution, as it determines how the material will behave, and can be used to explain the superior properties of composites over conventional materials. The methods used for modelling stress distribution in composites range widely in nature and complexity. These notes are intended as a short introduction to micromechanics of heterogeneous materials and modelling techniques.

Keywords: composites, inclusions, Eshelby method

1.   Introduction

The physical properties of a material are strongly influenced by the microstructure, which is designed during processing. The amounts of constituent elements forming the material are in many cases known beforehand as for example the volume fraction of reinforcing phase in composite materials. However, the final architecture of a microstructure is controlled only to a limited extent, and on the microscale the geometrical arrangement of second-phase inclusions is a result of complex and interacting micromechanical processes rather than a design variable. This seems to create an obstacle in modeling the relation between microstructure and overall material properties.

R. de Borst and T. Sadowski (eds.) *Lecture Notes on Composite Materials – Current Topics and Achievements*

In order to overcome this difficulty it is customarily assumed that the geometrical features of a microstructure are randomly distributed without any precise explanation of what randomness means and how it can be quantified. On the other extreme lies the assumption that the dispersion of fillers is regular. This significantly simplifies calculations, especially with the finite element method, and provides satisfactory results so long as highly non-linear phenomena are not concerned. For strongly localized effects such as initiation of microcracks and plastic zones, which subsequently propagate at different scale lengths, the regularity assumption may lead to seriously erroneous results. Therefore, it is important to describe dispersion characteristics of the microstructure in an unambiguous way by quantitative factors, which eventually could be related to physical properties of the material.

A multiphase material is by definition heterogeneous and its local properties vary spatially. If the material is statistically homogeneous, which means that the local material properties are constant when averaged over a representative volume element, then it is possible to replace the real disordered material by a homogeneous one in which the local material properties are the averages over the representative volume element in the real material. Estimation of those averages presents a fundamental issue for different effective medium theories, [Aboudi:91; Dvorak:92; Chen:92; Nemat-Nasser:93]. Morphological descriptors of microstructures and their dispersion characteristics play a fundamental role in determining property values of composites, [Pyrz:94a; Pyrz:98]. This is particularly true for highly non-linear phenomena such as fracture, where the microfailure threshold is dominated by extreme fluctuations of the stress field, and these load hot spots are strongly influenced by a local disorder of the reinforcement, [Axelsen:95; Ostoja:94; Pyrz:94]. When we examine the microstructure of a material we are looking at a very small sample of the structure. From this limited view we have tried to understand how the properties of the material relate to microstructure. The approach must necessarily be statistical in particular with respect to the measurement of the microstructure. A selection of representative volume elements where we seek to find microstructure-property relations must be based upon sound sampling techniques, [Cruz-Orive:81; Gundersen:87; Gundersen:99]. The size of representative volume elements must be such that overall properties of a heterogeneous material are obtained by some volume or ensemble-averaging procedure over the representative volume element, the length scale of which is much larger than the characteristic length scale of inclusions, but smaller than the characteristic length scale of a macroscopic sample. This description of the representative volume element must recognize the statistical nature of the microstructure of actual composites, that is, to include a sampling of all possible microstructural configurations that may occur in the composite. Each microstructure configuration gives an estimate for the overall quantity studied, and the average of estimates should give accurate results with the scatter of the estimates as small as possible. This is a quite difficult task, and an alternative approach would be to require only that the average is accurate and the scatter left unconstrained. [Drugan:96] showed that for a composite with elastic properties and spherical identical inclusions the minimum size of the representative volume element is surprisingly small. The effective elastic constants, calculated from ensemble averages of stress and strain field at finite lengths, converge quickly to those at an infinite length, i.e. the overall ones. For a large class of constituents the accuracy of the effective elastic constants derived at a length equal to two sphere diameters were within a few percent as compared to overall elastic constants calculated at infinite length. Gusev in [Gusev:97] took another route to studying periodic elastic composite with a disordered unit cell made up of a random dispersion of identical spheres. It was shown that already with a few dozen spheres in the unit cell, the scatter of the individual estimates obtained with the same number of spheres was small and the averages were practically stationary. Thus only a few realizations can reasonably predict the overall elastic constants of the periodic composite. Those results are valid for a particular family of composites and are aimed at the overall elastic properties. For other property considerations such as microcracking or fracture propagation, the size of the representative volume element must be related to the correlation length, which is the maximum length over which the fluctuations attributable to the individual defects are correlated, [Stauffer:92].

2. Representative Volume Element

Continuum mechanics deals with idealized materials consisting of material points and material neighbourhoods. It assumes that the material distribution, the stresses, and the strains within an infinitesimal material neighbourhood of a typical particle (or a material element) can be regarded as essentially uniform. On the microscale, however, the infinitesimal material neighbourhood, in general, is not uniform, consisting of various constituents with differing properties and shapes, i.e. an infinitesimal material element has its own complex and, in general, evolving microstructure. Hence, the stress and strain fields within the material element likewise are not uniform at the microscale level. One of the main objectives of micromechanics is to express in a systematic and rigorous manner the continuum quantities associated with an infinitesimal material neighbourhood in terms of the parameters that characterize the microstructure and properties of the micro-constituents of the material neighbourhood.

Figure 1 shows a continuum, and identifies a typical material point Q surrounded by an infinitesimal material element. When the macro-element is magnified, it may have its own complex microstructure consisting of very large number micro-heterogeneities such as grains, inclusions, voids, cracks, and

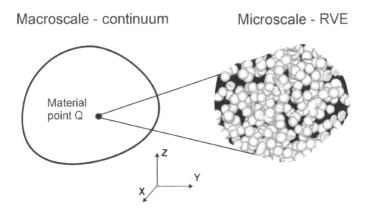

Figure 1. Representative volume element

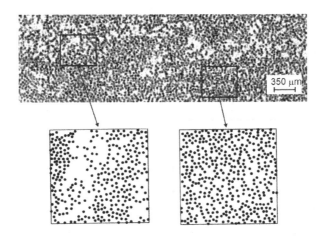

Figure 2. Cross section of unidirectional glass fibres/epoxy composite from X-ray microto-mographic scanning. Inserts show post-processed areas

other similar defects. To be representative, this RVE must include a large number of such micro-heterogeneities.

Figure 2 shows X-ray microtomographic image of the cross section in unidirectional glass fibres/epoxy composite. The matrix rich areas are clearly visible as a white spots. A fibre adjacent to the matrix rich area will be stressed very differently than fibre closely surrounded by neighbouring fibres. Therefore, the matrix rich areas constitute the smallest entity of the microstructure and are judged to have pronounced effects on the overall response of the continuum. The representative volume element is expected to have the same effective properties that the whole material. Thus, the typical dimension of the RVE, D, must be much larger than the average size of matrix rich areas, d; i.e. $D/d \gg 1$.

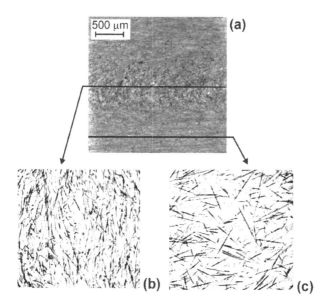

Figure 3. X-ray projection micrograph of short glass fibres/polyester composite (**a**) and reconstructed cross sections (**b**) and (**c**)

Figures 3(a–c) are micrographs of a polyester composite with short glass fibres. From the X-ray projection image in Fig. 3a, the microtomographic reconstruction has been performed to obtain cross section images in the core, (Fig. 3b), and outer layers, (Fig. 3c), of the composite sample. The composite was processed by injection molding, and different orientation distribution of fibres in these two sections is clear. A mechanism involves better orientation of fibres in the skin layers due to larger flow shearing as compared to the core. Thus, the variation of fibres orientation distribution at different sites of the sample can by itself induce nonhomogeneous stress field, apart from the disturbances caused by the presence of fibres. It then follows that the fibres orientation distribution has a pronounced influence on the local stiffness of the composite. The stiffness of the core layer may become as much as 30% less than the stiffness of outer layers. This example indicates that the microstructure is not statistically homogeneous, an assumption being a cornerstone behind the concept of representative volume element. It means that the volume-averaged characteristics of the composite, as described by the RVE, do not correspond to the ensemble characteristics, i.e. characteristics obtained from different microstructural configurations that occur in the composite.

Spatial dispersion of fillers or inclusions is orchestrated during the manufacturing process and depends on the desired volume fraction of these entities, their shapes, sizes and orientations. These three morphological factors are inherently related, especially for the volume fractions approaching the maximum packing fraction achievable for a given shape, size and orientation. Morphology is the study of form. The form of an object is the shape and structure of that object as distinguished from its material components. The structure may include dispersion and orientation distribution of features of the object. The shape or form of an object is not a well-defined concept. The notion of a geometrical structure or texture is not purely objective. The concept of an inclusion size is inherently connected to the notion of shape. Convex inclusions with geometrically regular shapes can, in principle, be assigned with linear sizes. For non-convex inclusions and inclusions with irregular shapes, other measures of size such as surface area or volume must be used. The concept of size becomes critical if one wants to compare inclusions of unequal shape. In this situation statement defining a size for one type of shapes not necessarily holds true for another shape of inclusions. The boundary between computational geometry and morphology is therefore rather fuzzy.

Another important question is what constitutes an underlying essential microconstituent. This is also a relative concept depending on the particular problem and the particular objective. It must be addressed through systematic microstructural observation at the level of interest, and must be guided by experimental results. Perhaps one of the most vital decisions is the definition of the RVE. An optimum choice would be one that includes the most dominant features that have first-order influence on the overall properties of interest, and at the same time yields the simplest model. This can be done only through a coordinated sequence of microscopic (small-scale) and macroscopic (continuum-scale) observation, experimentation and analysis. In many problems in the mechanics of materials, suitable choices often emerge naturally in the course of the examination of the corresponding physical attributes and the experimental results.

3.   The Eshelby Equivalent Inclusion Method

Internal stresses are commonplace in almost any material which is mechanically inhomogeneous. Typically, their magnitude varies according to the degree of inhomogeneity for an externally loaded polycrystalline cubic metal, differently oriented crystallites will be stressed to different extents, but these differences are usually quite small. For a composite consisting of two distinct constituents with different stiffnesses, these disparities in stress will commonly be much larger. Internal stresses arise as a result of some kind of misfit between the shapes of the constituents (matrix and reinforcement, i.e. fibre, whisker or particle). Such a misfit could arise from a temperature change, but a closely related situation is created during mechanical loading: a stiff inclusion tends to deform less than the surrounding matrix. Analysis of the stress required to mate up the inclusion and matrix across the interface allows the prediction of

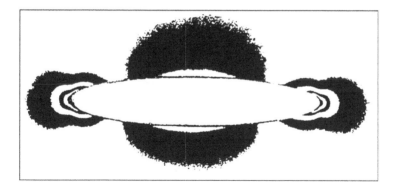

Figure 4. Photoelastic fringes from ellipsoidal inclusion subject to axial compression

properties such as thermal expansivity and stiffness. For an arbitrary inclusion shape, this analysis can be carried out only numerically, but for the special case of an ellipsoid an analytical technique can be employed. The key point here is that the ellipsoid, which can have any aspect ratio, has a uniform stress at all points within it.

Figure 4 shows the slice from an axisymmetric three-dimensional photoelastic model containing a prolate ellipsoidal inclusion of stiffness ∼2.5 times that of the matrix, viewed between crossed polars. The model was loaded in axial compression. The pattern of fringes (contours of equal principal stress difference) shows that the stress within the matrix fluctuates in a complicated manner, the stress being largest at the inclusion ends and smaller than the applied stress about the equator. The stress within the inclusion, on the other hand, is larger than the applied stress and is uniform throughout.

The Eshelby technique is based on representing the actual inclusion by one made of matrix material which has an appropriate misfit strain, so that the stress field is the same as for the actual inclusion (see [Markenscoff:06]). Suppose a region within a homogeneous medium were suddenly to transform in shape, so that it no longer fitted freely into the hole in the matrix from which it came. What would the stress field look like? The answer to this question would at first sight appear to have little to do with calculating the stresses within composites, but Eshelby showed that there is an elegant solution to this problem, which can be applied to a wide variety of other situations. The consequences of a spontaneous transformation of the type discussed above can best be visualized in terms of displacement maps (Fig. 5).

In these diagrams, the grid lines represent the displacement of an originally square mesh, while the thickness of lines represents the stiffness. A transformation (i.e. a shape change) imposed on a region within a matrix tends to cause complex distortions in both the transformed region and the surrounding matrix. This holds for non-uniform strains (for a linear change in width

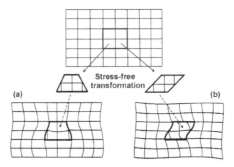

Figure 5. Local distorsions caused by (**a**) non-uniform strains and (**b**) uniform strains

with height, Fig. 5(a)) and uniform strains (for a simple shear, Fig. 5(b)). It is clear that the elastic strain field is very complicated, both inside and outside the constrained transformed region, and for this reason an analytic solution is not usually possible. An example of such a spontaneous shape change is provided by a martensitic transformation. However, when the transformed region is ellipsoidal in shape and the shape change is a uniform one, the mathematics become tractable. This is because under these conditions the stress and strain within the enclosed phase are uniform (see Fig. 4).

3.1   Average Strain and Stress Theorems

Under conditions of an imposed macroscopically homogeneous stress or deformation field on the RVE, the average stress and strain are representatively defined by

$$< \varepsilon_{ij} > = \frac{1}{V} \int_V \varepsilon_{ij} \, dV \qquad (1a)$$

$$< \sigma_{ij} > = \frac{1}{V} \int_V \sigma_{ij} \, dV \qquad (1b)$$

where V is the volume of the RVE and the symbol $<>$ denotes volume averaging. Homogeneous boundary conditions applied on the surface of a homogeneous body will produce a homogeneous field there. Such boundary conditions are obtained by imposing displacements at the boundary S in the form

$$u_i(S) = \varepsilon_{ij}^0 x_j \qquad (2)$$

where ε_{ij}^0 are constant strains. Alternatively, traction can be imposed on S so that

$$t_i = \sigma_{ij}^0 n_j \qquad (3)$$

where σ_{ij}^0 are constant stresses and **n** is the unit outward normal vector to S. To calculate the average strains in composite material it appears that one must solve the elasticity problem of the RVE subjected to the displacement homogeneous boundary conditions, Eq. (2). The strain–displacement relations are

$$\varepsilon_{ij} = \frac{1}{2}(u_{i,j} + u_{j,i}) \tag{4}$$

Substituting (4) in (1a) yields

$$2V < \varepsilon_{ij} >= \int_{V_1} (u_{i,j}^{(1)} + u_{j,i}^{(1)}) \, dV_1 + \int_{V_2} (u_{i,j}^{(2)} + u_{j,i}^{(2)}) \, dV_2 \tag{5}$$

where "1" and "2" denote phase 1 and 2 of the two phased composites with V_1, V_2 being the volumes occupied by the two phases. The use of Gauss' theorem

$$\int_V u_{i,p} \, dV = \int_S u_i \, n_p \, dS \tag{6}$$

implies that

$$2V < \varepsilon_{ij} >= \int_{S_1} (u_i^{(1)} n_j + u_j^{(1)} n_i) \, dS + \int_{S_2} (u_i^{(2)} n_j + u_j^{(2)} n_i) \, dS \tag{7}$$

where S_1 and S_2 are the bounding surfaces of phases 1 and 2, respectively. The surfaces S_1 and S_2 contain the interfaces S_{12} and the external surface S. For perfect bonding between the phases, i.e.

$$u_i^{(1)} = u_i^{(2)} \quad \text{on} \quad S_{12}$$

it follows that the contributions from S_{12} in the two integrals in Eq. (7) cancel each other. This leads to

$$< \varepsilon_{ij} >= \frac{1}{2V} \int_S (u_i n_j + u_j n_i) \, dS \tag{8}$$

Substituting in Eq. (8) the homogeneous boundary conditions (2) and again using Gauss' theorem (6) yields

$$< \varepsilon_{ij} >= \frac{1}{2V} \int_S (\varepsilon_{ip}^0 x_p n_j + \varepsilon_{jp}^0 x_p n_i) \, dS =$$

$$= \frac{1}{2V} \varepsilon_{ij}^0 \int_V 2x_{p,i} \, dV = \varepsilon_{ij}^0 \tag{9}$$

where $x_{p,i} = \delta_{pi}$, i.e. the volume averaged strain within RVE is equal to the constant strain applied on the surface.

The homogeneous boundary conditions with traction applied on S, produce a stress field in the composite whose average, $< \sigma_{ij} >$, is identical to the constant stress $\sigma_{ij}{}^0$. To this end, consider the equilibrium equations in the absence of body forces

$$\sigma_{ij,j} = 0 \tag{10}$$

which implies that

$$(\sigma_{ik}x_j)_k = \sigma_{ik,k}x_j + \sigma_{ij}x_{j,k} = \sigma_{ij}\delta_{jk} = \sigma_{ij} \tag{11}$$

Substituting this relation in (1a) provides

$$V < \sigma_{ij} >= \int_V (\sigma_{ik}x_j)_k \, dV$$

and by Gauss' theorem we have

$$V < \sigma_{ij} >= \int_{S_1} \sigma_{ik}^{(1)} x_j n_k^{(1)} ds + \int_{S_2} \sigma_{ik}^{(2)} x_j n_k^{(2)} ds \tag{12}$$

Since tractions are continuous at the interfaces S_{12}, i.e.

$$\sigma_{ij}^{(1)} n_j^{(1)} = -\sigma_{ij}^{(2)} n_j^{(2)} \quad \text{on} \quad S_{12}$$

the contributions from S_{12} to the two integrals cancel each other and Eq. (12) reduces to

$$V < \sigma_{ij} >= \int_S \sigma_{ik}x_j n_k ds = \sigma_{ij}^{(0)} \int_S x_j n_k dS$$
$$= \sigma_{ik}^{(0)} \int_V x_{jk} \, dV = V\sigma_{ik}^{(0)} \tag{13}$$

Thus

$$< \sigma_{ij} >= \sigma_{ij}^{(0)}$$

i.e. the volume average stress within RVE is equal to constant stress applied at the boundary.

3.2   Relation Between Averages

Let us consider the homogeneous boundary conditions (2) according to which displacements are applied on the surface S of a RVE. For a two-phase composite with perfect bonding between the constituents

$$< \varepsilon_{ij} >= c_1 < \varepsilon_{ij} >^{(1)} + c_2 < \varepsilon_{ij} >^{(2)} \tag{14}$$

where c_α with $\alpha = 1, 2$ denote the volume fractions of the phases, and

$$< \varepsilon_{ij} >^{(\alpha)} = \frac{1}{V_\alpha} \int_{V_\alpha} \varepsilon_{ij}^{(\alpha)} \, dV \qquad (15)$$

which is the average strain in the phase α. Similarly,

$$< \sigma_{ij} >^{(\alpha)} = \frac{1}{V_\alpha} \int_{V_\alpha} \sigma_{ij}^{(\alpha)} \, dV \qquad (16)$$

and

$$< \sigma_{ij} >= c_1 < \sigma_{ij} >^{(1)} + c_2 < \sigma_{ij} >^{(2)} \qquad (17)$$

By the average strain theorem we have

$$< \varepsilon_{ij} >= \varepsilon_{ij}^{(0)}$$

and the constitutive law in the phases is

$$< \sigma_{ij} >^{(\alpha)} = C_{ijkl}^{(\alpha)} < \varepsilon_{kl} >^{(\alpha)} \qquad \alpha = 1, 2 \qquad (18)$$

where $C_{ijkl}^{(\alpha)}$ are stiffnesses of phases.

The constitutive law for a composite can be written in terms of averages

$$< \sigma_{ij} >= C_{ijkl}^* < \varepsilon_{ij} > \qquad (19)$$

where C_{ijkl}^* is as yet unknown stiffness of the composite. From Eqs. (9), (14), (16), (17) and (18) substituted to Eq. (19) one gets

$$C_{ijkl}^* < \varepsilon_{ij} >=< \sigma_{ij} >$$
$$C_{ijkl}^* \varepsilon_{kl}^0 = c_1 < \sigma_{ij} >^{(1)} + c_2 < \sigma_{ij} >^{(2)}$$
$$C_{ijkl}^* \varepsilon_{kl}^0 = c_1 C_{ijkl}^{(1)} < \varepsilon_{kl} >^{(1)} + c_2 C_{ijkl}^{(2)} < \varepsilon_{kl} >^{(2)} \qquad (20)$$
$$C_{ijkl}^* \varepsilon_{kl}^0 = C_{ijkl}^{(1)} (< \varepsilon_{kl} > -c_2 < \varepsilon_{kl} >^{(2)}) + c_2 C_{ijkl}^{(2)} < \varepsilon_{kl} >^{(2)}$$
$$C_{ijkl}^* \varepsilon_{kl}^0 = C_{ijkl}^{(1)} \varepsilon_{kl}^0 + c_2 (C_{ijkl}^{(2)} - C_{ijkl}^{(1)}) < \varepsilon_{kl} >^{(2)}$$

Equation (20) implies that the effective moduli can be determined from the elastic moduli of the phases provided the average strain $< \varepsilon_{kl} >^{(2)}$ in the inclusion phase 2 is known.

3.3 The Eshelby Solution

The case where a given material is infinitely extended is of particular interest for the mathematical simplicity of the solution as well as for its practical importance. When the solution is applied to inclusion problems, it can be assumed with sufficient accuracy that the materials are infinitely extended, since the size of the inclusions is relatively small compared to the size of the macroscopic material samples.

Definitions:

- Eigenstrain: nonelastic strains such as thermal expansion, phase transformation, initial strains.

- Eigenstress: self-equilibrated internal stress caused by one or several of these eigenstrains in bodies which are free from any other external force and surface constraints.

The term "residual stresses" have been frequently used for the self-equilibrated internal stresses when they remain in the material after fabrication or plastic deformation. Eigenstresses are called thermal stresses when thermal expansion is a cause of the corresponding elastic field. For example, when a part Ω of a material has its temperature raised by T, thermal stress σ_{ij} is induced in D by the constraint from the part which surrounds Ω, (Fig. 6). The thermal expansion αT, where α is the linear thermal expansion coefficient, constitutes the thermal expansion strain

$$\varepsilon_{ij}^* = \alpha T \, \delta_{ij}$$

where δ_{ij} is the Kronecker delta. The thermal expansion strain caused when Ω can be expanded freely with the removal of the constraint from the surrounding part.

The actual strain is then the sum of the thermal stress by Hooke's law. The thermal expansion strain is a typical example of an eigenstrain. In the elastic theory of eigenstrains and eigenstresses, however, it is not necessary to attribute ε_{ij}^* to any specific source. The source could be phase transformation, precipitation, plastic deformation or a fictitious source necessary for the equivalent inclusion method.

When an eigenstrain ε_{ij}^* is prescribed in a finite subdomain Ω in a homogeneous material D, and it is zero in the matrix D-Ω, then Ω is called an **inclusion**. The elastic moduli of the inclusion Ω and the matrix d-Ω are the **same**.

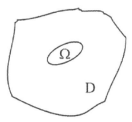

Figure 6. Inclusion Ω

If a subdomain Ω in a material D has elastic moduli **different** from those of the matrix D-Ω, then Ω is called an **inhomogeneity**. Applied stresses will be disturbed by the existence of the inhomogeneity. This disturbed stress field will be simulated by an eigenstress field by considering a fictitious eigenstrain $\varepsilon_{ij}{}^*$ in Ω in a homogeneous material.

The major contribution made by Eshelby was to establish the relation

$$\varepsilon_{ij}{}^{(2)\text{total}} = S_{ijkl}\,\varepsilon_{kl}{}^* \quad \text{in} \quad \Omega \tag{21}$$

where the total strain in Ω is given by the sum of the eigenstrain and the resulting elastic strain, i.e.

$$\varepsilon_{ij}^{(2)total} = \varepsilon_{ij}^{(2)} + \varepsilon_{ij}^* \tag{22}$$

The eigenstrain $\varepsilon_{ij}{}^*$ is uniform in Ω and the resulting strain (22) is also uniform in Ω. This is the most valuable result from the Eshelby solution. (The development of Eq. (21) is omitted here). In the expression (21) S_{ijkl} is called the Eshelby tensor; it has the following properties:

- It is symmetric with respect to the first two indices and the second two indices

$$S_{ijkl} = S_{jikl} = S_{ijlk}$$

however, it is not, in general, symmetric with respect to the exchange of ij and kl, i.e.

$$S_{ijkl} \neq S_{klij}$$

- It is independent of the material properties of the inclusion Ω.

- It is completely defined in terms of the aspect ratios of the ellipsoidal inclusion Ω and the elastic parameters of the surrounding matrix $D - \Omega$.

- When the surrounding matrix $D - \Omega$ is isotropic, then S_{ijkl} depends only on the Poisson ratio of the matrix and the aspect ratios of Ω.

If an ellipsoidal inclusion has its principal half axes denoted by a_1, a_2 and a_3, then the Eshelby tensor components for special shapes of inclusions are as follows:

Sphere ($a_1 = a_2 = a_3$)

$$S_{1122} = S_{2233} = S_{1133} = S_{1133} = S_{2211} = S_{3322} = \frac{5\nu - 1}{15(1 - \nu)}$$

$$S_{1212} = S_{2323} = S_{3131} = \frac{4 - 5\nu}{15(1 - \nu)}$$

$$S_{1111} = S_{2222} = S_{3333} = \frac{7 - 5\nu}{15(1 - \nu)}$$

Elliptic cylinder ($a_3 \to \infty$)

$$S_{1111} = \frac{1}{2(1-\nu)} \left[\frac{a_2^2 + 2a_1a_2}{(a_1 + a_2)} + (1 - 2\nu)\frac{a_2}{a_1 + a_2} \right]$$

$$S_{2222} = \frac{1}{2(1-\nu)} \left[\frac{a_1^2 + 2a_1a_2}{a_1 + a_2} + (1 - 2\nu)\frac{a_1}{(a_1 + a_2)} \right]$$

$$S_{3333} = 0$$

$$S_{1122} = \frac{1}{2(1-\nu)} \left[\frac{a_2^2}{(a_1 + a_2)} - (1 - 2\nu)\frac{a_2}{a_1 + a_2} \right]$$

$$S_{2233} = \frac{1}{2(1-\nu)}\frac{2\nu a_1}{a_1 + a_2}$$

$$S_{3311} = 0$$

$$S_{1133} = \frac{1}{2(1-\nu)}\frac{2\nu a_2}{a_1 + a_2};$$

$$S_{2211} = \frac{1}{2(1-\nu)} \left[\frac{a_1^2}{(a_1 + a_2)^2} - (1 - 2\nu)\frac{a_1}{a_1 + a_2} \right]$$

$$S_{3322} = 0; \quad S_{1212} = \frac{1}{2(1-\nu)} \left[\frac{a_1^2 + a_2^2}{2(a_1 + a_2)^2} + \frac{(1 - 2\nu)}{2} \right]$$

$$S_{2323} = \frac{a_1}{2(a_1 + a_2)}; \quad S_{3131} = \frac{a_2}{2(a_1 + a_2)}$$

Penny shape ($a_1 = a_2 >> a_3$)

$$S_{1111} = S_{2222} = \frac{13 - 8\nu}{32(1-\nu)}\pi\frac{a_3}{a_1}; \quad S_{3333} = 1 - \frac{1 - 2\nu}{1 - \nu}\frac{\pi}{4}\frac{a_3}{a_1}$$

$$S_{1122} = S_{2211} = \frac{8\nu - 1}{32(1-\nu)}\pi\frac{a_3}{a_1}; \quad S_{1133} = S_{2233} = \frac{2\nu - 1}{8(1-\nu)}\pi\frac{a_3}{a_1}$$

$$S_{3311} = S_{3322} = \frac{\nu}{1 - \nu}\left(1 - \frac{4\nu + 1}{8\nu}\pi\frac{a_3}{a_1}\right)$$

$$S_{1212} = \frac{7 - 8\nu}{32(1-\nu)}\pi\frac{a_3}{a_1}; \quad S_{1313} = S_{2323} = \frac{1}{2}\left(1 + \frac{\nu - 2}{1 - \nu}\frac{\pi}{4}\frac{a_3}{a_1}\right)$$

when $a_3 \to \infty$

$$S_{2323} = S_{3131} = \frac{1}{2}; \quad S_{3311} = S_{3322} = \frac{\nu}{1 - \nu}; \quad S_{3333} = 1,$$

and all other $S_{ijkl} = 0$.

3.4 Equivalent Inclusion Method

Consider an infinitely extended material with the elastic moduli $C_{ijkl}^{(1)}$ containing an ellipsoidal inhomogeneity with the elastic moduli $C_{ijkl}^{(2)}$. We investigate the disturbance in an applied stress caused by the presence of this inhomogeneity. Let us denote the applied stress at infinity by σ_{ij}^0 and the corresponding strain ε_{ij}^0. The stress disturbance and the strain disturbance are denoted by $\sigma_{ij}^{(\alpha)}$ and $\varepsilon_{ij}^{(\alpha)}$, $\alpha = 1, 2$, respectively. The total stress (actual stress) is $\sigma_{ij}^0 + \sigma_{ij}^{(1)}$ in the matrix and $\sigma_{ij}^0 + \sigma_{ij}^{(2)}$ in the inhomogeneity. The total strains are $\varepsilon_{ij}^0 + \varepsilon_{ij}^{(1)}$ and $\varepsilon_{ij}^0 + \varepsilon_{ij}^{(2)}$, respectively. Hooke's law is written

$$
\begin{aligned}
\sigma_{ij}^0 + \sigma_{ij}^{(2)} &= C_{ijkl}^{(2)}(\varepsilon_{kl}^0 + \varepsilon_{kl}^{(2)}) \quad \text{in} \quad \Omega \\
\sigma_{ij}^0 + \sigma_{ij}^{(1)} &= C_{ijkl}^{(1)}(\varepsilon_{kl}^0 + \varepsilon_{kl}^{(1)}) \quad \text{in} \quad D - \Omega
\end{aligned}
\tag{23}
$$

The equivalent inclusion method is used to simulate the stress disturbance using the eigenstress resulting from an inclusion which occupies the space Ω. Consider an infinitely extended homogeneous material with the elastic moduli $C_{ijkl}^{(1)}$ everywhere, containing domain Ω with an eigenstrain ε_{ij}^*, (Fig. 7). The eigenstrain has been introduced here arbitrarily in order to simulate the inhomogeneity problem using the inclusion problem.

When this homogeneous material is subjected to the applied strain ε_{ij}^0 at infinity, the resulting total stress in the inclusion is

$$
\sigma_{ij}^0 + \sigma_{ij}^{(2)} = C_{ijkl}^{(1)}(\varepsilon_{kl}^0 + \varepsilon_{kl}^{(2)total} - \varepsilon_{kl}^*) \quad \text{in} \quad \Omega
\tag{24}
$$

The necessary and sufficient condition for the equivalency of the stresses and strains in these two problems of the inhomogeneity and inclusion, is an identity between Eqs. $(23)_1$ and (24), i.e.

$$
C_{ijkl}^{(2)}(\varepsilon_{kl}^0 + \varepsilon_{kl}^{(2)}) = C_{ijkl}^{(1)}(\varepsilon_{kl}^0 + \varepsilon_{kl}^{(2)total} - \varepsilon_{kl}^*)
\tag{25}
$$

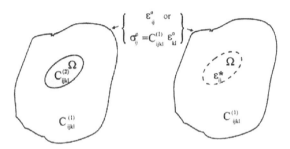

Figure 7. Equivalent inclusion method

where $\varepsilon_{kl}^{(2)\text{total}} - \varepsilon_{kl}^{*} = \varepsilon_{kl}^{(2)}$ is the elastic strain in the inclusion, Eq. (22). For the inhomogeneity problem (i.e. left hand side of Eq. (25)) the strain $\varepsilon_{kl}^{(2)}$ is equal to the total strain $\varepsilon_{kl}^{(2)\text{total}}$, then from Eqs. (25) and (21) one gets

$$C_{ijk}^{(2)}(\varepsilon_{kl}^0 + S_{klmn}\varepsilon_{mn}^*) = C_{ijkl}^{(1)}(\varepsilon_{kl}^0 + S_{klmn}\varepsilon_{mn}^* - \varepsilon_{kl}^*) \qquad (26)$$

Knowing elastic constants of the inhomogeneity and the matrix together with the Eshelby tensor, we can find from Eq. (26) components of the eigenstrain ε_{ij}^{*} in terms of the remote strain ε_{ij}^{0}. After obtaining ε_{ij}^{*} the stress in the inhomogeneity, we can find $\sigma_{ij}^{0} + \sigma_{ij}^{(2)}$ from Eq. (23)$_1$ or Eq. (24).

In an RVE there is a unique dependence between the average strains in the phases and the overall strain in the composite. Let this be written as

$$<\varepsilon_{ij}>^{(1)} = A_{ijkl}^{(1)} <\varepsilon_{kl}> = A_{ijkl}^{(1)}\varepsilon_{kl}^0$$
$$<\varepsilon_{ij}>^{(2)} = A_{ijkl}^{(2)}\varepsilon_{kl}^0 \qquad (27)$$

with

$$c_1 A_{ijkl}^{(1)} + c_2 A_{ijkl}^{(2)} = I_{ijkl}$$

where I_{ijkl} is the unit tensor. $A_{ijkl}^{(\alpha)}$, $\alpha = 1, 2$, are called concentration tensors. Then, substituting the average strain in the inclusion phase 2, i.e. Eq. (27)$_2$ into the expression (20), the required effective stiffness of the composite can be obtained as

$$C_{ijkl}^{*} = C_{ijkl}^{(1)} + c_2(C_{ijmn}^{(2)} - C_{ijmn}^{(1)}) A_{mnkl}^{(2)} \qquad (28)$$

From Eq. (28) it is clear that the knowledge of the concentration tensor is sufficient to determine the effective stiffness of a composite, provided that the constituent properties and volume fractions of phases are known. Since this is a tremendously difficult task, different models have been introduced to approximate the concentration tensor.

The simplest model follows from the Eshelby concept and is called the "dilute" approximation: the concentration tensor is approximated by embedding a single particle in an all-matrix material. In the "dilute" approximation the volume fraction of particles is small enough that each single particle does not "see or feel" its neighbours, and therefore they may be considered independently. Since the distances between particles are large as compared to dimensions of the particles, we may consider each particle embedded in an infinite matrix. Then from Eq. (27)$_2$

$$\underline{A}^{(2)} = \frac{\underline{\varepsilon}^{(2)}}{\underline{\varepsilon}^0} \qquad (29)$$

The strain in the inhomogeneity

$$\underline{\varepsilon}^{(2)} = \underline{\varepsilon}^{(2)total} + \underline{\varepsilon}^0 \qquad (30)$$

or

$$\underline{\varepsilon}^{(2)} = \underline{S}\,\underline{\varepsilon}^* + \underline{\varepsilon}^0 \qquad (31)$$

It follows from Eqs. (25) and (31) that

$$\underline{C}^{(1)}(\underline{\varepsilon}^{(2)} - \underline{\varepsilon}^*) = \underline{C}^{(2)}\underline{\varepsilon}^{(2)}$$

or

$$\underline{C}^{(1)}\underline{\varepsilon}^* = (\underline{C}^{(1)} - \underline{C}^{(2)})\underline{\varepsilon}^{(2)} \qquad (32)$$

From the Eshelby solution (21) the eigenstrain can be calculated as

$$\underline{\varepsilon}^* = \underline{S}^{-1}\underline{\varepsilon}^{(2)total}$$

or

$$\underline{\varepsilon}^* = \underline{S}^{-1}(\underline{\varepsilon}^{(2)} - \underline{\varepsilon}^0) \qquad (33)$$

where $\underline{S}^{-1}$ is an inverse Eshelby tensor. Then substituting (33) into (32) yields

$$\underline{C}^{(1)}[\underline{S}^{-1}(\underline{\varepsilon}^{(2)} - \underline{\varepsilon}^0)] = (\underline{C}^{(1)} - \underline{C}^{(2)})\underline{\varepsilon}^{(2)}$$

or

$$[\underline{C}^{(1)}\underline{S}^{-1} - (\underline{C}^{(1)} - \underline{C}^{(2)})]\,\underline{\varepsilon}^{(2)} = \underline{C}^{(1)}\,\underline{S}^{-1}\underline{\varepsilon}^0 \qquad (34)$$

Multiplication by $\underline{S}[\underline{C}^{(1)}]^{-1}$ provides

$$\{\underline{I} - \underline{S}[\underline{C}^{(1)}]^{-1}(\underline{C}^{(1)} - \underline{C}^{(2)})\}\underline{\varepsilon}^{(2)} = \underline{\varepsilon}^0 \qquad (35)$$

It follows that the requested concentration tensor of an elliptical inclusion, embedded in a matrix subjected to uniform deformation at large distances from the inclusion, is given by

$$\underline{A}^{(2)} = \frac{\underline{\varepsilon}^{(2)}}{\underline{\varepsilon}^0} = \{\underline{I} - \underline{S}[\underline{C}^{(1)}]^{-1}(\underline{C}^{(1)} - \underline{C}^{(2)})\}^{-1} \qquad (36)$$

and the effective stiffness can be determined from Eq. (28).

Typical results are shown in Fig. 8, which gives axial and transverse stiffness prediction for aligned short fibre composites where the short fibre was modelled as an alongated ellipsoid with the aspect ratio of $a_3/a_1 = a_2$ varying from 1 to 100. Figure 8(a) confirms that in the practicable volume fraction range up to about 40–50%, fibres with fairly high aspect ratios are needed in order to effect substantial improvements in the stiffness. The transverse stiffness predictions in Fig. 8(b), however, show clearly that the aspect ratio has very little effect on the transverse stiffness. This is the case for all composites.

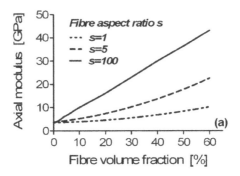

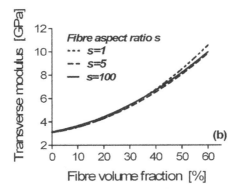

Figure 8. The Young's modulus of short glass fibres/epoxy composite as a function of fibre volume fraction for (**a**) axial and (**b**) transverse loading

4. The Mori–Tanaka Theory

An equivalent inclusion method provides a first-order approximation, yields an explicit result for a composite stiffness $\underline{C}^*$, but neglects particle interaction. It is therefore valid only at small concentration of particles. A better approximation, which takes into account particle interactions, is the Mori–Tanaka theory. This method assumes that the average strain $< \underline{\varepsilon} >^{(2)}$ in the interacting inhomogeneities can be approximated by that of a single inhomogeneity embedded in an infinite matrix subjected to the uniform **average matrix strain** $< \underline{\varepsilon} >^{(1)}$. This is illustrated in Fig. 9.

The problem to be solved according to this model is that of a single inhomogeneity in a certain large volume V' which is enclosed by a surface S', and subjected to the boundary condition

$$u_p(S') = < \varepsilon_{pq} >^{(1)} x_q \tag{37}$$

The solution of this problem is

$$< \varepsilon_{mn} >^{(2)} = T_{mnpq} < \varepsilon_{pq} >^{(1)} \tag{38}$$

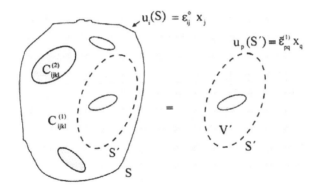

Figure 9. Schematic representation of the Mori–Tanaka model

where $\underline{T}$ is determined from the solution of a single inhomogeneity embedded in an infinite matrix subjected to boundary conditions (37). Substituting (38) into expression (14) and remembering the statement of the average strain theorem (Eq. (9)) we find

$$\underline{\varepsilon}^0 = c_1 < \varepsilon >^{(1)} + c_2 \underline{T} < \varepsilon >^{(1)}$$

or

$$< \underline{\varepsilon} >^{(1)} = (c_1 \underline{I} + c_2 \underline{T})^{-1} \varepsilon^0 \qquad (39)$$

Substituting Eq. (39) into (38) leads to the following relation

$$\underline{A}^{(2)} = \frac{< \varepsilon >^{(2)}}{\varepsilon^0} = \underline{T}(c_1 \underline{I} + c_2 \underline{T})^{-1} \qquad (40)$$

Thus, the concentration tensor $\underline{A}^{(2)}$, which relates an average strain in the inhomogeneity to the uniform remote strain field, is related to the concentration tensor $\underline{T}$ via Eq. (40). Consequently, the effective stiffness of the composite is given from Eqs. (28) and (40) as

$$\underline{C}^* = \underline{C}^{(1)} + c_2(\underline{C}^{(2)} - \underline{C}^{(1)})\underline{T}(c_1 \underline{I} + c_2 \underline{T})^{-1} \qquad (41)$$

The concentration tensor $\underline{T}$ is determined from the solution of a single inhomogeneity embedded in an infinite matrix subjected to boundary conditions (37). From the Eshelby equivalent inclusion method, Eq. (26), we have

$$\underline{C}^{(2)}(< \varepsilon >^{(1)} + \underline{S}\varepsilon^*) = \underline{C}^{(1)}(< \varepsilon >^{(1)} + \underline{S}\varepsilon^* - \varepsilon^*) \qquad (42)$$

where $\underline{\varepsilon}^0$ has been replaced by the average matrix strain $< \varepsilon >^{(1)}$ since now it constitutes the homogeneous boundary conditions. For the inhomogeneity the strain is given by

$$< \underline{\varepsilon} >^{(2)} = \underline{\varepsilon}^{(2)total} + < \varepsilon >^{(1)} = \underline{S}\varepsilon^* + < \varepsilon >^{(1)} \qquad (43)$$

The expression (43) is analogous to (31) with an appropriate replacement of the remote strain field. Then, calculating the eigenstrain from (43), i.e.

$$\underline{\varepsilon}^* = \underline{S}^{-1}(<\underline{\varepsilon}>^{(2)} - <\underline{\varepsilon}>^{(1)})$$

and substituting to Eq. (42) with (43) in mind yields

$$\underline{C}^{(2)} <\underline{\varepsilon}>^{(2)} = \underline{C}^{(1)}(<\underline{\varepsilon}>^{(2)} - \underline{\varepsilon}^*)$$
$$\Downarrow$$
$$\underline{C}^{(1)}\underline{\varepsilon}^* = (\underline{C}^{(1)} - \underline{C}^{(2)}) <\underline{\varepsilon}>^{(2)}$$
$$\Downarrow \tag{44}$$
$$\underline{C}^{(1)}\underline{S}^{-1}(<\underline{\varepsilon}>^{(2)} - <\underline{\varepsilon}>^{(1)}) = (\underline{C}^{(1)} - \underline{C}^{(2)}) <\underline{\varepsilon}>^{(2)}$$
$$\Downarrow$$
$$[\underline{C}^{(1)}\underline{S}^{-1} - (\underline{C}^{(1)} - \underline{C}^{(2)})] <\underline{\varepsilon}>^{(2)} = \underline{C}^{(1)}\underline{S}^{(1)} <\underline{\varepsilon}>^{(1)}$$

Multiplication of both sides of the Eq. (44) by a factor $\underline{S}^{-1}(\underline{C}^{(1)})^{-1}$ yields

$$[\underline{I} - \underline{S}(\underline{C}^{(1)})^{-1}(\underline{C}^{(1)} - \underline{C}^{(2)})] <\underline{\varepsilon}>^{(2)} = <\underline{\varepsilon}>^{(1)}$$

and recalling the definition of the concentration tensor T, Eq. (38) one obtains

$$\underline{T} = [\underline{I} - \underline{S}(\underline{C}^{(1)})^{-1}(\underline{C}^{(1)} - \underline{C}^{(2)})]^{-1} \tag{45}$$

Then, finally, the effective stiffness is determined from Eq. (41) with the substitution of the expression (45).

The engineering constants of the composite can be derived from the stiffness tensor $\underline{C}^*$ (Eq. (41)) which is too elaborate for manual manipulations and must be evaluated with a computer programme. As an example, the components of the stiffness tensor $\underline{C}^*$ have been calculated for a short fibre composite with perfectly aligned fibres modelled as an elliptic cylinder with equal semi-axis $a_1 = a_2$. The numerical values of the stiffness components are as follows

$$[C_{ijkl}] = \begin{bmatrix} 7.450 & 3.428 & 3.179 & 0 & 0 & 0 \\ & 7.450 & 3.179 & 0 & 0 & 0 \\ & & 0.295 & 0 & 0 & 0 \\ & symmetric & & 3.082 & 0 & 0 \\ & & & & 3.082 & 0 \\ & & & & & 1.590 \end{bmatrix} \times 10^{-10}[GPa]$$

The stiffness has been calculated for a nylon PA6/glass fibre composite with 35% volume fraction of fibres. The matrix and the fibre stiffness were, respectively, 3.0 [GPa] and 73.5 [GPa], whereas the Poisson ratio of the matrix was $\nu = 0.33$. From the stiffness components one can extract engineering constants of the composite, and the Young's modulus of the composite along the direction of fibres' alignment reads E = 27.68 [GPa], whereas the transverse modulus, i.e. along the cross section perpendicular to the fibres privilege direction is $E_t = 5.77$ [GPa].

References

J. Aboudi, *Mechanics of Composite Materials*, Elsevier, Amsterdam, 1991.

M. Axelsen and R. Pyrz, in *Proceedings of the IUTAM Symposium on Microstructure Property Interactions in Composite Materials*, ed. R. Pyrz, Kluwer, The Netherlands, 1995, pp. 15–26.

T. Chen., G.J. Dvorak and Y. Benveniste, *J. Comp. Mat.*, 1992, **59**, 539–546.

L.M. Cruz-Orive and E.R. Weibel, *J. Microsc.*, 1981, **122**, 235–257.

W.J. Drugan and J.R. Willis, *J. Mech. Phys. Solids*, 1996, **44**, 497–524.

G.J. Dvorak, *Proc. R. Soc. Lon. A*, 1992, **437**, 311–327.

H.J.G. Gundersen and E.B. Jensen, *J. Microsc.*, 1987, **187**, 229–263.

H.J.G. Gundersen, E.B. Jensen, K. Kiêu and J. Nielsen, *J. Microsc.*, 1999, **199**, 199–211.

A.A. Gusev, *J. Mech. Phys. Solids*, 1997, **45**, 1449–1459.

X. Markenscoff and A. Gupta, (Eds.) *Collected Works of J.D. Eshelby – The Mechanics of Defects and Inhomogeneities,* Springer, 2006

S. Nemat-Nasser and M. Hori, *Micromechanics: Overall Properties of Heterogeneous Materials*, North-Holland, Amsterdam, 1993.

M. Ostoja-Starzewski, P.Y. Sheng and I. Jasiuk, *ASME J. of Engng. Mat. Tech.*, 1994, **116,** 384–391.

R. Pyrz, *Comp. Sci. Techn.*, 1994 (a), **50**, 539–546.

R. Pyrz and B. Bochenek, *Sci. Engng. Comp. Mat.*, 1994, **3**, 95–109.

R. Pyrz and B. Bochenek, *Int. J. Solids Struct.*, 1998, **35**, 2413–2427.

D. Stauffer and A. Aharoni, *Introduction to Percolation Theory*, 2nd edition, Taylor & Francis, Washington DC, 1992.

NON-SYMMETRIC THERMAL SHOCK
IN CERAMIC MATRIX COMPOSITE (CMC)
MATERIALS

Tomasz Sadowski
Department of Solid Mechanics
Lublin University of Technology
Nadbystrzycka 40
20-618 Lublin
Poland
t.sadowski@pollub.pl

Abstract   A methodology has been proposed to describe CMC and functionally graded materials (FGM) thermomechanical response subjected to sudden changes of the temperature field. Appropriate gradation of the composite properties can significantly improve thermal shock response of the material itself or a structural element. The governing equations for the temperature field under transient thermal loading are formulated, and solution of the analytical and two numerical methods FEA and FD were discussed. The role of the thermal residual stress is significant in the analysis of thermal shock problems. In a non-symmetrical gradation profile they create initial curvature of the FGM structural element. The basic fracture mechanics idea is presented and is applied to CMC and FGM which were subjected to a transient temperature field. In particular the crack-bridging mechanism plays an important role in the assessment of the composites response. Numerical examples dealing with 1-D and 2-D non-symmetric thermal shock problem illustrate the effectiveness of the theoretical methods to the solution of practical problems.

Keywords: composites, non-symmetric thermal shock, crack propagation

1.   Introduction to ceramic and metal matrix composites

Current engineering requires the application of new materials with improved properties. This is particularly important in aerospace engineering, high temperature applications, microelectronics, etc., where new composite materials

R. de Borst and T. Sadowski (eds.) *Lecture Notes on Composite Materials – Current Topics and Achievements*

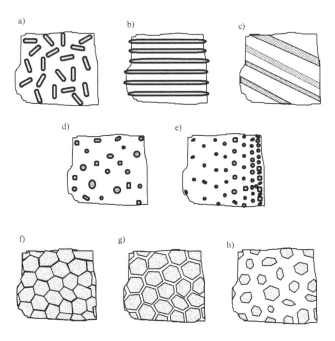

Figure 1. Type of composites: (**a**) short-fiber, (**b**) long fiber, (**c**) layered, (**d**) particle, (**e**) FGM (functionally gradient material), (**f**) polycrystalline with different fracture properties of grain boundary, (**g**) polycrystalline with small interfaces, (**h**) polycrystalline with thick interfaces

play the basic role in any innovation. One can classify two types of composites used for thermal applications: ceramic matrix composites (CMC) and metal matrix composites (MMC). The production of CMCs and MMCs with controlled microstructures allows us to obtain a material with controlled thermomechanical properties, which are necessary for a particular purpose, e.g.: Evans (1990), Munz and Fett (1999) and Sadowski (2005). Figure 1 shows typical internal microstructures in two-phase components.

Generally, CMCs and MMCs can be treated as multiphase materials, where components made of different materials are joined during technological processes (e.g. pressing and sintering) in order to create a new and improved material. Ceramic composites are used particularly in high temperature applications. The typical example is conventional thermal barrier coating (TBC), which is a thin one layer film (Fig. 2a), or a multi-layer structure (Fig. 2b) created on a metallic substrate using a plasma-spraying technique. Another classical example is a layered composite with different compositions in each layer (stepwise functionally graded TBC) (Fig. 3). However TBCs in a form of layered materials exhibit poor interfacial bonding and high residual and thermal stress between constituents, and low toughness, which can lead to easy cracking or spallation. These disadvantages can be partially removed by the

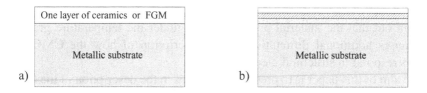

Figure 2. Thermal barrier coating: (**a**) one layer TBC, (**b**) multi-layer TBC

Al_2O_3
Al_2O_3 + 5 wt% ZrO_2
Al_2O_3 + 10 wt% ZrO_2
Al_2O_3 + 15 wt% ZrO_2
Al_2O_3 + 20 wt% ZrO_2

Figure 3. Stepwise functionally graded TBC

introduction of a two-phase composite in the form of functionally graded materials (FGM), which is made of two materials of different natures: ceramic and metal, (e.g. Erdogan, 1995; Neubrand and Rödel, 1997). The ceramics in an FGM will cause thermal barrier effects and will protect the metal from corrosion and oxidation. On the other hand the metallic component causes toughening and strengthening of the FGMs. The idea of FGMs relies on the continuous variation of volume contents of both components in such a way that the resulting non-uniform microstructure in the material exhibits the desired thermomechanical properties at the macroscale. Application of FGMs in TBCs improve bonding strength and also reduces residual as well as thermal stresses in bonded dissimilar materials. It also improves wear resistant layers in friction linings, brake linings, gears, bearings, etc.

The modeling of modern CMCs and MMCs material requires the co-operation of scientists and industrial workers. Fundamental studies on the mechanical behaviour of CMCs and MMCs comprise investigations concerning estimation of their thermomechanical properties, taking into account current microstructure and the properties of micro-constituents (e.g. Nemat-Nasser and Hori, 1993; Pyrz, 1994; and this volume Pyrz and Bochenek 1994, 1998; Munz and Fett, 1999; Moon et al., 2005; Sadowski et al., 2005a,b, 2006b, 2007c, 2008; Sadowski and Samborski, 2003a,b). The important problems are these: the strength of CMCs as related to component properties, yield strength and fracture toughness. These investigations should be advanced both in theoretical and experimental aspects.

The mechanical response of CMCs and MMCs should be analysed for mechanical and thermal aspects. Mechanical loading: monotonic, variable, fatigue and creep will characterize the composites' behaviour under room temperature. Thermal loading will comprise the following cases: a steady state under elevated (up to 1100°C), high temperature (at level of 1600°C) or non-stationary temperature, thermal shock and thermal fatigue. It is necessary to point out that the material properties vary under the temperature gradient; this causes additional difficulties in the description of how the CMCs and MMCs respond in structural elements.

The aim of this paper is to present basic ideas in the description of the CMCs behaviour under non-symmetric and transient temperature states, including assessment of thermomechanical properties and crack growth processes resulting from technological surface defects. In particular 1-D and 2-D thermal shock problems in FGMs and layered ceramic composites will be discussed.

2.   Thermomechanical properties of CMCs and MMCs

The importance of non-symmetric thermal shock problems arises from many practical applications, e.g. TBC in combustion engines, turbines, etc. Therefore an evaluation of the CMC and MMC properties during thermal shock is essential for its reliable use in high-temperature structures, e.g. Suresh and Mortensen (1998), Zuiker (1995), Noda (1999), Jeulin and Ostoja-Starzewski (2001), Ostoja-Starzewski and Schulte (1996), Ostoja-Starzewski et al. (1994), Jiang et al. (2000), Kim and Noda (2002), Sadowski et al. (2005, 2006a, 2007a) and Postek et al. (2005).

The initial composition and its morphology have a basic influence on the behaviour of modern CMCs and MMCs. In relation to this data one can classify the considered materials in the following manner:

1. Homogeneous composites, containing uniform distribution of all phase components,

2. Heterogeneous composites, which can be treated as statistically homogeneous materials (Torquato, 1998), i.e. it is possible to specify part of the material microstructure within the Representative Volume Element (RVE), whose averages are constant and correspond to volume averages of the specimen

3. Randomly heterogeneous composites, which are statistically inhomogeneous and the material microstructure fluctuation is totally random (Ostoja-Starzewski, 2007). Averages over the RVE in this case depend on local position x_i in the heterogeneous material

Another classification of the composites with the specific initial geometry of microstructure include:

1. Discontinuously reinforced composites (DRC), in which each reinforcing particle is totally surrounded by a matrix phase

2. Interpenetrating phase composites (IPC), where phases of the CMCs and MMCs interpenetrate in arbitrary or controlled (FGM) ways

The basic problems in the characterization of CMC composites are estimations of:

- Elastic properties

- Thermal properties

- Crack resistance properties

and for CMCs (containing plastic particles) and MMCs:

- Yield stress

- Stress–strain behaviour under the yield stress, i.e. tangent modulus in case of hardening composites

Here, we limit our consideration to two-dimensional composites made of matrix (ceramic phase – C or C1) and reinforcement (ceramic phase – C2 or metallic one – M).

The basic problem concerning the description of the CMCs or MMCs behaviour is the assessment of elastic material properties. Several approaches can be potentially applicable:

- Isostrain (Voigt) and isostress (Reuss) approximation models (e.g. Ravichandran, 1994; Tuchinskii 1983)

- Equivalent inclusion method (Pyrz, this volume) which is valid only for a small concentration of particles (e.g. Eshelby, 1957)

- The Mori–Tanaka model, which takes into account particle interaction (Mori and Tanaka, 1973; Benveniste, 1987; Reiter et al., 1997; Pyrz – this volume)

- Self-consistent methods (SC): classical (e.g. Hill, 1965; Willis, 1977; Reiter et al., 1997) and three-phase self-consistent method (e.g. Wilkinson et al., 2001)

- The Hashin–Shtrikman method (Hashin and Shtrikman, 1963) is derived from a variational procedure for linear elastic composites with a statistically isotropic microstructure (an assembly of random discontinuous spherical inclusions). Generalisation for nonlinear composites was done by Talbot and Willis (1985, 1992, 1994, 2004), Ponte Castañeda (1991); Ponte Castañeda and Sequet (1998), Willis (2000), Fleck and Willis (2004), Aifantis and Willis (2005)

- Effective medium approximation methods – EMA – (e.g. Kreher and Pompe, 1989, Moon et al., 2005), which consider an assembly of random contacting elliptical inclusions assembly; Ostoja-Starzewski (2007)

- Model of inhomogeneous distribution of fully penetrable spheres (Tor-quato, 1998, 2002) for analysis of disordered heterogeneous media or random nonlinear two-phase media (Aifantis and Willis, 2006)

- Topological approach for random distribution of inclusions (Fan et al. 1992, 1993) by characterizing the microstructure in terms of topological parameters such as: contiguity and continuous volume

- Numerical models – finite element analysis (FEA) approaches for:

    1. Homogeneous materials – with a unit cell model which, with application of appropriate boundary conditions, represent a periodically repeating 2-D or 3-D structure (e.g. Aboudi, 1991, Wegner and Gibson, 2000)

    2. Heterogeneous materials – with a unit cell model which is created by

        - Finite-length inclusions arranged in a periodic manner in one direction and arbitrary in the second and third (e.g. 2-D higher-order theory, Pindera et al., 1998, 2002)

        - The image analysis of real heterogeneous materials, which reconstructs random media (an inverse problem), e.g. Sadowski et al. (2005, 2006b, 2007c); Sadowski and Nowicki (2008)

2.1 Two-phase CMC with different elastic components

For the purpose of heat transfer modelling, the thermal and elastic properties of the specimens were evaluated in the simplest way, according to the linear rule of mixture (RoM – isostress and isotrains approximations), e.g. Miller (1969), Tamura et al. (1973), Noda (1999), and Sadowski et al. (2007a). Any property of the two-phase CMC material $\bar{P}(x_i, T)$, related to 3-D space $\{x_i\}$ and temperature T (in [°K]), can be estimated as having the properties of its components, i.e. the first phase (ceramics C1) $P^{C1}(T)$ and the second phase (ceramics C2) $P^{C2}(T)$ as well as their appropriate volume contents: $V^{C1} = (1 - V^{C2})$ and V^{C2}. Thus, in the simplest case of CMCs, the thermal or elastic properties with constant volume content $V^{C2} = \text{const.}$ can be calculated through the relation:

$$\bar{P}(T) = P^{C1}(T)[1 - V^{C2}] + P^{C2}(T)V^{C2} \qquad (1)$$

For example, layered CMCs, with a different volume content of two phases in each layer, were considered in Sadowski et al. (2007a). The CMCs were created from two kinds of ceramics: Al_2O_3 and ZrO_2 (Fig. 3). For calculation of the thermal conductivity of a single, ith layer of considered CMCs, the following formula was used

$$^{(i)}\bar{k} = ^{(i)}k^{Al_2O_3/ZrO_2} = k^{Al_2O_3}(1 - V_{(i)}^{ZrO_2}) + k^{ZrO_2}V_{(i)}^{ZrO_2} \quad (2)$$

For FGM materials, the properties of CMCs also depend on position $\{x_i\}$ in the composite structure and we have

$$\bar{P}(x_i, T) = P^{C1}(T)[1 - V^{C2}(x_i)] + P^{C2}(T)V^{C2}(x_i), \quad (3)$$

where $V^{C2}(x_i)$ is the volume content of the ceramics C2.

2.2   Two-phase CMCs with plastic inclusions and MMCs with elastic inclusions

2.2.1   Thermomechanical properties

The more general class of materials are the composites produced from elastic and plastic phases. Application of RoM gives a formula similar to (1) and (3), i.e.:

$$\bar{P}(T) = P^C(T)[1 - V^M] + P^M(T)V^M, \quad (4)$$

where $P^C(T)$ and $P^M(T)$ are the properties of the ceramic and metallic phases. The appropriate volume contents of both phases are equal to $V^C = (1 - V^M)$ and V^M. For example, in further analysis, the following denotation will be used for Young's moduli: $\bar{E}$ – for CMCs and MMCs, E^C – for ceramic matrix and E^M for the metallic phase.

An example of this type of composites is Al_2O_3/Al (e.g. Moon et al., 2005), WC/Cu (e.g. Schmauder et al., 1999), WC/Co (e.g. Dong and Schmauder, 1996; Li et al., 1999; Sadowski and Nowicki, 2008; Felten et al., 2008), ZrO_2/Ti-6Al-4V (Tanigawa et al., 1997), Al_2O_3/Ni (e.g. Ravichandran, 1995; Giannakopoulos et al., 1995; Pitakthapanaphong and Busso, 2002; Kouzeli and Dunand, 2003), etc.

An interesting discussion of the currently used models for the prediction of the effective Young's modulus of elastic–plastic composites at room temperature was presented by Moon et al. (2005). According to this paper, up till now, the theoretical efforts to assess elastic behaviour over the whole composition range ($V^M = 0.05 \div 0.97$) have had limited success. For CMC and MMC materials with a fixed composite structure, modelling of elastic properties is sufficiently accurate by a homogenisation technique (unit cell method), e.g. Suresh and Mortensen (1998), Ravichandran (1994), and Tuchinskii (1983). However, for composites with a random structure (e.g. IPC) they suggest the application of EMA, particularly for materials having Young's moduli components ratio $E^C/E^M \in (3.7 \div 5.6)$. Moreover, in this approach it is necessary to introduce an appropriate microstructural shape factor. For Al_2O_3/Al, composite EMA is the most consistent method for assessment of $\bar{E}$ over the entire composition range, $V^M = 0.05 \div 0.97$. On the other hand, Pitakthapanaphong and Busso (2002) suggest the application of a self-consistent approach for modelling the Al_2O_3/Ni composite, where the mismatch of Young's moduli of 2 phases is relatively low ($E^{Al_2O_3}/E^{Ni} = 1.8$). However, for composites with a large elastic property mismatch between phases, the better application is a variational approach (e.g. Li and Ponte Castañeda, 1994; Ponte Castañeda and Sequet 1998). With the higher values of V^M, composite materials can be classified as MMCs and treated as ductile solids reinforced by different shaped elastic inclusions, e.g. Al/Al_2O_3 composite – Kouzeli and Dunand (2003).

In all these methods, only the approximations of the real material structure can be analysed. Experimental data gives the final answer as to the correct value of Young's modulus of CMCs. The combination of the SEM observations, including computer tomography, allows for creation of RVE which exhibits the real material structure. Application of FEA, creates a promising tool for assessment of mechanical properties of WC/Co, e.g. Sadowski and Nowicki (2008) for 2-D case.

The estimation of the material properties for FGMs ceramic/metal CMCs is a more difficult task. Generally, we can extend (5) by introduction of the dependence of thermomechanical features on the reference co-ordinate system $\{x_i\}$:

$$\bar{P}(x_i, T) = P^C(T)[1 - V^M(x_i)] + P^M(T)V^M(x_i) \qquad (5)$$

2.2.2   Modelling of the whole stress–strain curve for MMC

The theoretical estimation of the whole stress–strain curves for ceramic/metal MMCs is a more difficult task, particularly so for FGMs.

An interesting approach in the characterisation of the whole stress–strain curve for a WC/Co composite was proposed by Tamura et al. (1973), using the intermediate law of mixture. In this so called TTO model, the stress–strain curve was predicted by adopting J_2 flow theory, using an RoM approach in order to evaluate the material properties: elastic constants, yield stress and tangent modulus. The extension of this model was made in Jin et al. (2003) to include J_2 flow theory with isotropic hardening.

In the TTO model, the stress $\bar{\sigma}$ and $\bar{\varepsilon}$ in the composite, which is subjected to uniaxial tension, can be calculated as corresponding average uniaxial stresses and strains of both composite components

$$\bar{\sigma} = \sigma^C[1 - V^M] + \sigma^M V^M \qquad (6)$$

and

$$\bar{\varepsilon} = \varepsilon^C[1 - V^M] + \varepsilon^M V^M, \qquad (7)$$

where σ^C, σ^M, ε^C, ε^M denote the average stresses and strains in both phases. In the TTO model the additional parameter of stress to strain ratio is introduced in the form:

$$q = (\sigma^C - \sigma^M)/|\varepsilon^C - \varepsilon^M|. \qquad (8)$$

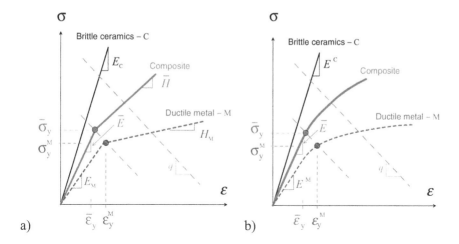

Figure 4. Stress–strain curve for elastic–plastic MMC: (**a**) with linear hardening, (**b**) with non-linear hardening

The value of q belongs to the interval $(0, \infty)$. For $q = 0$, the two phases have the same stress level, whereas for $q \to \infty$ ceramics and metal exhibit the same deformations in the loading direction. In reality for any MMCs q has a nonzero finite value. Figure 4 shows schematically the behaviour of both MMC components

$$\sigma^C = E^C \varepsilon^C, \qquad \sigma^M = E^M \varepsilon^M \qquad (9)$$

and the total MMCs response. Young's modulus can be obtained using (6)–(9)

$$\bar{E} = \left[E^C(1 - V^M) + E^M V^M \frac{q + E^C}{q + E^M} \right] \Big/ \left[(1 - V^M) + V^M \frac{q + E^C}{q + E^M} \right]. \qquad (10)$$

Poisson's ratio $\bar{\nu}$ is calculated according to the RoM

$$\bar{\nu} = \nu^C [1 - V^M] + \nu^M V^M. \qquad (11)$$

The TTO model assumes that the MMC yields once the metal constituents yield ($\sigma^M = \sigma_y^M$). Knowing the yield stress σ_y^M, Young's moduli: E^C, E^M, q and volume content V^M, we can express metal phase of the composite $\bar{\sigma}_y$:

$$\bar{\sigma}_y(V^M) = \sigma_y^M \left[(1 - V^M) \frac{q + E^M}{q + E^C} \frac{E^C}{E^M} + V^M \right] \qquad (12)$$

If the metallic phase behaves as elastic–plastic material with linear hardening modulus H^M, then MMC also follows a similar response, and the linear hardening modulus $\bar{H}$ is

$$\bar{H}^{\mathrm{M}}(V^{\mathrm{M}}) = \left[(1 - V^{\mathrm{M}})E^{\mathrm{C}} + V^{\mathrm{M}}H^{\mathrm{M}}\frac{q + E^{\mathrm{C}}}{q + H^{\mathrm{M}}}\right] \Big/ \left[V^{\mathrm{M}}\frac{q + E^{\mathrm{C}}}{q + H^{\mathrm{M}}} + (1 - V^{\mathrm{M}})\right]$$

(13)

For power-law behaviour for metal, (Jin et al., 2003) (Fig. 4b), the constitutive relations for metal phase and MMCs are as follows:

$$\varepsilon^{\mathrm{M}} = \varepsilon_{\mathrm{y}}^{\mathrm{M}}\left(\frac{\sigma^{\mathrm{M}}}{\sigma_{\mathrm{y}}^{\mathrm{M}}}\right)^{\mathrm{n}^{\mathrm{M}}}, \qquad \bar{\varepsilon} = \bar{\varepsilon}_{\mathrm{y}}\left(\frac{\bar{\sigma}}{\bar{\sigma}_{\mathrm{y}}}\right)^{\bar{\mathrm{n}}}$$

(14)

where $\varepsilon_{\mathrm{y}}^{\mathrm{M}} = \sigma_{\mathrm{y}}^{\mathrm{M}}/E^{\mathrm{M}}$ and $\bar{\varepsilon}_{\mathrm{y}} = \bar{\sigma}_{\mathrm{y}}/\bar{E}$. n^{M} and $\bar{\mathrm{n}}$ are the hardening exponents of the metal and MMC. Using (6)–(9) and (14) one can obtain the following parametric equations describing the composite response:

$$\frac{\bar{\varepsilon}}{\bar{\varepsilon}_{\mathrm{y}}} = \frac{(1 - V^{\mathrm{M}})\bar{E}}{q + E^{\mathrm{C}}}\frac{\sigma^{\mathrm{M}}}{\bar{\sigma}_{\mathrm{y}}} + \frac{q + V^{\mathrm{M}}E^{\mathrm{C}}}{q + E^{\mathrm{C}}}\frac{\bar{E}}{E^{\mathrm{M}}}\frac{\sigma_{\mathrm{y}}^{\mathrm{M}}}{\bar{\sigma}_{\mathrm{y}}}\left(\frac{\sigma^{\mathrm{M}}}{\sigma_{\mathrm{y}}^{\mathrm{M}}}\right)^{\mathrm{n}_{\mathrm{M}}}$$

(15)

$$\frac{\bar{\sigma}}{\bar{\sigma}_{\mathrm{y}}} = \frac{V^{\mathrm{M}}q + E^{\mathrm{C}}}{q + E^{\mathrm{C}}}\frac{\sigma^{\mathrm{M}}}{\bar{\sigma}_{\mathrm{y}}} + \frac{(1 - V^{\mathrm{M}})qE^{\mathrm{C}}}{(q + E^{\mathrm{C}})E^{\mathrm{M}}}\frac{\sigma_{\mathrm{y}}^{\mathrm{M}}}{\bar{\sigma}_{\mathrm{y}}}\left(\frac{\sigma^{\mathrm{M}}}{\sigma_{\mathrm{y}}^{\mathrm{M}}}\right)^{\mathrm{n}_{\mathrm{M}}}$$

(16)

For metal material $V^{\mathrm{M}} = 1$, (15) takes the form $(14)_1$. It is important to point out that the stress–strain curve determined according to (15) and (16) does not follow the power function (15). The particular application of these theories to model Ti/TiB FGM composite was presented by Jin et al. (2003).

These formulations obey the simple phenomenological modelling of composite behaviour in application to thermal shock problems. However, the formulation of the constitutive equations for CMCs and MMCs are a more complicated problem, because it is necessary to take into account the real microstructure (e.g. architecture of reinforcement in composite) of the material (Werner et al., 1994; Sequet, 1997; Moulinec and Sequet, 2003; Sadowski, 2005; Ostoja-Starzewski, 1994, 2007; Dong and Schmauder, 1996; Pindera et al., 1998; Li et al., 1999). For example, Sadowski and Nowicki (2008) have estimated elastic properties $\bar{E}$, $\bar{\nu}$, the yield strength $\bar{\sigma}_{\mathrm{y}}$ and the whole stress-strain curve for the WC/Co composite considering RVE and standard FEA. The analysis of the RVE response in uniaxial loading was the best application of so called "generalized plane strain conditions" with straight edges imposed on the RVE. The stress–strain correlation was very close to a bi-linear model (Fig. 4a).

2.2.3 Constitutive equations for FGMs in 3-D formulation by self-consistent approach

Let us consider a three-layered model of the structure (Fig. 5) (e.g. Giannakopoulos et al., 1995; Pitakthapanaphong and Busso, 2002). In the

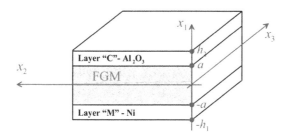

Figure 5.   Three-layered model of the FGM structure

middle FGM layer, the volume contents of ceramic is denoted by V^C and metallic phase by V^M. In order to describe the thermo-elastoplastic response of the FGM material one can apply a self-consistent approach, Pitakthapanaphong and Busso, (2002). The constitutive equation in the case of 3-D FGM materials, can be expressed as

$$
\begin{aligned}
\dot{\bar{\sigma}}_{ij}(x_1) = {} & 2\bar{\mu}(x_1)\left\{\dot{\bar{\varepsilon}}_{ij}(x_1) - \dot{\bar{\varepsilon}}^{\mathrm{P}}_{ij}(x_1)\right\} + \delta_{ij}\bar{\lambda}(x_1)\left\{\dot{\bar{\varepsilon}}_{kk}(x_1) - \dot{\bar{\varepsilon}}^{\mathrm{P}}_{kk}(x_1)\right\} \\
& - \delta_{ij}\left\{2\bar{\mu}(x_1) + 3\bar{\lambda}(x_1)\right\}\bar{\alpha}(x_1)\dot{T}
\end{aligned}
\tag{17}
$$

where $\dot{\bar{\sigma}}_{ij}$ and $\dot{\bar{\varepsilon}}_{ij}$ are the averaged macroscopic stress and strain rate components in the FGM. $\dot{\bar{\varepsilon}}^{\mathrm{P}}_{ij}$ is the overall plastic strain rate. $\dot{T} = \partial T/\partial t$ is the temperature rate. The elastic material properties $\bar{\mu}(x_1)$, $\bar{\lambda}(x_1)$ are expressed as functions of Young's modulus and Poisson's ratio:

$$
\begin{aligned}
\bar{E}(x_1) &= E^{\mathrm{M}} + (E^{\mathrm{C}} - E^{\mathrm{M}})V^C(x_1) \\
\bar{\nu}(x_1) &= \nu^{\mathrm{M}} + (\nu^{\mathrm{C}} - \nu^{\mathrm{M}})V^C(x_1)
\end{aligned}
\tag{18}
$$

and thermal properties are described by

$$
\bar{\alpha}(x_1) = \alpha^{\mathrm{M}} + (\alpha^{\mathrm{C}} - \alpha^{\mathrm{M}})V^C(x_1)
\tag{19}
$$

The plastic strains in FGMs are equal to

$$
\dot{\bar{\varepsilon}}^{\mathrm{P}}_{ij} = V^{\mathrm{M}}\dot{\varepsilon}^{\mathrm{P(M)}}_{ij}
\tag{20}
$$

where $\dot{\varepsilon}^{\mathrm{P(M)}}_{ij}$ is the plastic strain rate in the metallic phase. Denoting by $\sigma'^{\mathrm{(M)}}_{ij}$ the deviator stress component and by $\sigma^{\mathrm{M}}_e = \sqrt{3/2(\sigma'^{\mathrm{(M)}}_{ij}\sigma'^{\mathrm{(M)}}_{ij})}$ the equivalent stress in the phase M, we can calculate $\dot{\varepsilon}^{\mathrm{P(M)}}_{ij}$ from

$$
\dot{\varepsilon}^{\mathrm{P(M)}}_{ij} = \left\langle \dot{\chi}^{\mathrm{M}} \right\rangle \frac{\sigma'^{\mathrm{(M)}}_{ij}}{\sigma^{\mathrm{M}}_e}
\tag{21}
$$

In this relation $\dot{\chi}^{\mathrm{M}}$ is equal to

$$\langle \dot{\chi}^{\mathrm{M}} \rangle = \begin{cases} (3/2)(\dot{\sigma}_{\mathrm{e}}^{\mathrm{M}}/H^{\mathrm{M}}) & \text{for loading} \quad \sigma_{\mathrm{e}}^{\mathrm{M}} \geq \sigma_{\mathrm{y}}^{\mathrm{M}} \\ 0 & \text{for } \sigma_{\mathrm{e}}^{\mathrm{M}} \geq \sigma_{\mathrm{y}}^{\mathrm{M}} \text{ and unloading } \sigma_{\mathrm{e}}^{\mathrm{M}} = \sigma_{\mathrm{y}}^{\mathrm{M}} \end{cases}$$

(22)

here H^{M} is the hardening modulus of the phase M, whereas $\sigma_{\mathrm{y}}^{\mathrm{M}}$ – its yield limit.

The micro average stress rate in elasto-plastic phase FGM is defined by

$$\dot{\sigma}_{ij}^{(\mathrm{M})} = \dot{\bar{\sigma}}_{ij} + 2\bar{\mu}(1 - \beta)\left\{ V^{\mathrm{M}}\dot{S}_{ij}^{(\mathrm{M})} - \dot{S}_{ij}^{(\mathrm{M})} \right\}$$

(23)

where β is Eshelby's elastic accommodation factor, and $\dot{S}_{ij}^{(\mathrm{M})}$ is the evolution of the interphase accommodation variable.

This formula predicts the behaviour of the elastoplastic MMC for composites with a medium elastic property mismatch between phases M and C. For Al$_2$O$_3$/Ni FGM (Fig. 5) Young's moduli ratio was equal to $E^{\mathrm{Al_2O_3}}/E^{\mathrm{Ni}} = 1.775$.

However, for composites with a large elastic property mismatch between phases $E^{\mathrm{C}}/E^{\mathrm{M}}$, the better application is a variational approach (e.g. Li and Ponte Castañeda, 1994; Ponte Castañeda and Sequet, 1998).

3.   Temperature field under transient thermal loading

Because of the very complicated internal structure of CMC or MMC materials, an analytical solution of the transient temperature field is a complicated mathematical problem. Therefore, besides the exact analytical solutions, numerical approaches with applications of finite difference (FD) and FEA methods are applied in order to establish a temperature field inside the composite.

In the most general case, let us assume a solid material occupying the space V which is surrounded by a surface S in the 3-D co-ordinate system $\{x_i\}$ ($i = 1, 2, 3$) (Fig. 6). Suppose the temperature varies from point to point and from time to time continuously with respect to $\{x_i\}$ and time t, which is denoted by $T(x_i, t)$ the transient temperature field.

The basic equations governing the transient thermal loading in the elastic anisotropic solids are as follows:

- Fourier's law:

$$q_i = -k_{ij}^T{}_{,j},$$

(24)

where q_i are the components of the heat flux vector $\vec{q}(x_i, t)$, k_{ij} are components of the thermal conductivity tensor $k(x_i, t)$ of the CMCs composite and $T_{,j}(x_i, t) = \partial T/\partial x_j$ is the temperature gradient.

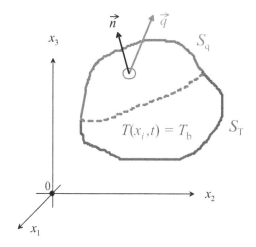

Figure 6. Boundary conditions in heat transfer problems

- The heat flux equation:

$$q_{i,i} + Q = \rho\, c_v \frac{\partial T}{\partial t},$$ (25)

where Q is the internal heat generation rate per unit volume, $\partial T/\partial t$ is the time derivative of the temperature field $T(x_i, t)$. $\rho(x_i, t)$ and $c_v(x_i, t)$ are mass density and specific heat of the composite material.

By combining (24) and (25) one can get a basic equation describing the temperature diffusion during heat transfer:

$$\left(k_{ij} \frac{\partial T}{\partial x_j} \right)_{,i} = -Q + \rho\, c_v \frac{\partial T}{\partial t}$$ (26)

Equation (26) is the general formulation; all material properties are considered to be functions of space $\{x_i\}$ and time t and the thermal anisotropy of the solid material is included. In the simplest case the material properties could not vary with time t, i.e. $\rho(x_i)$, $c_v(x_i)$ and $\mathbf{k}(x_i)$. Moreover, the thermal conductivity tensor is generally considered to be symmetric, i.e. $k_{ij} = k_{ji}$. If the material is anisotropic, heat will not necessarily flow in the direction of the temperature gradient. For isotropic material the thermal conductivity is described by the scalar value $k = (1/3)\, k_{ij} \delta_{ij}$.

The thermal conductivity equation (26) must be solved under given

- Initial conditions, which specify temperature distribution for $t = 0$, $T(x_i, 0) = T_0(x_i, 0)$

- Heat boundary conduction as specified surface heat flow on S_q

$$q_i n_i = h_b$$ (27)

or in a form of specified temperature distribution on S_T

$$T(x_i, t) = T_b \qquad (28)$$

Here $S_q + S_T = S$, $\vec{n}$ is the unit vector normal to the exterior of S. Equation (27) specifies thermal flux, and h_b is the surface convective heat transfer coefficient at the boundary S_q : it should be known or determined by experiments (e.g. Sadowski et al., 2007a); its value is negative if it is directed towards the interior of the material. Similarly the function T_b should be known at the beginning of the heat conduction process.

The solution of diffusion equation (26) could be done

-   Analytically by application of the perturbation method (Noda, 1999), Laplace or Fourier transformations or by Green's functions approach (e.g. Kim and Noda, 2002)

-   Numerically by FEA or FD (e.g. Morton and Mayers, 2005; Wang et al., 2004; Wang and Mai, 2005; Sadowski et al., 2007a; Nakonieczny and Sadowski, 2008).

3.1   FEA approach for heat transfer equation (26)

Let us focus on a numerical approach to solve the heat transfer Eq. (26). Consider the variational formulation of the Eq. (26), supposing that the material or structural element is subjected to virtual temperature change δT. Assume that δT satisfies the boundary conditions (27) and (28). Then (26) after integration over the whole volume of the material, becomes

$$\int_V \left[\rho\, c_v \frac{\partial T}{\partial t} - \left(k_{ij} \frac{\partial T}{\partial x_j} \right)_{,i} - Q \right] \delta T \; dV = 0. \qquad (29)$$

Taking into account Fourier's law (25) and applying the Green formula one can get

$$\int_V \left[(\rho\, c_v \frac{\partial T}{\partial t} - Q)\delta T - (q_i \delta T_{,i}) \right] dV + \int_{S_q} h_b \delta T \; dS = 0. \qquad (30)$$

This equation could be solved by FEA. In order to do it let us divide the whole material into a finite number of elements connected at nodal points. Having the shape function matrix $[N_e]$ and a vector $\{T\}$ defining the temperature values at the nodal points, one can specify the temperature field $T(x_i, t)$ for each element "e"

$$T(x_i, t) = [N_e] \{T\}. \qquad (31)$$

Following (31), the temperature gradient can be expressed in the form

$$\left\{ \begin{array}{c} \frac{\partial T}{\partial x_1} \\ \frac{\partial T}{\partial x_2} \\ \frac{\partial T}{\partial x_3} \end{array} \right\} = [B_e]\{T\}; \quad \text{where} \quad [B_e] = [\nabla][N_e] \tag{32}$$

and $[\nabla(...)] = \left[\ \partial(...)/\partial x_1 \quad \partial(...)/\partial x_2 \quad \partial(...)/\partial x_3 \ \right]^T$ is a differential operator of the quantity $(...)$.

Fourier's law is

$$\{q\} = \left\{ \begin{array}{c} q_1 \\ q_2 \\ q_3 \end{array} \right\} = -[k_{ij}][B_e]\{T\}. \tag{33}$$

$[k_{ij}]$ is the thermal conductivity matrix.

Introducing (31)–(33) to (30) one can get the heat equation in the form

$$[C]\{\dot{T}\} + [K]\{T\} = \{\hat{q}\} \tag{34}$$

where $\{\dot{T}\}$ denotes the time derivative of T and

$$[C] = \int_\Omega \rho\, c_v [N_e]^T [N_e] \mathrm{d}\Omega, \tag{35}$$

$$[K] = \int_\Omega [B_e]^T [k][B_e] \mathrm{d}\Omega, \tag{36}$$

$$\{\hat{q}\} = -\int_S [N_e]^T h_b\, \mathrm{d}S + \int_\Omega Q[N_e]^T \mathrm{d}\Omega. \tag{37}$$

Ω is the space occupied by finite elements approximating the solid CMCs and S is the corresponding boundary surface.

There are some methods used in order to solve matrix differential equation (34). Wang and Mai (2005) have used direct numerical differences in time for this purpose.

3.2   FD approach for heat transfer equation (26)

Another way to solve the parabolic differential equation (34) is through the application of FD methods, explicit forward time centered space (FTCS) method (e.g. Sadowski et al., 2007a) or generalized ADI (alternating-direction implicit) finite difference scheme (Sadowski and Nakonieczny, 2008).

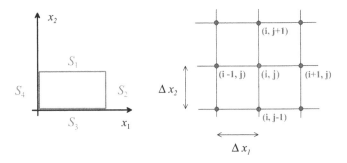

Figure 7. FD network for 2-D heat transfer problem

Let us consider a heat transfer problem when there is no internal heat sources, e.g. $Q = 0$. Then (26) becomes

$$k_{ij}\frac{\partial^2 T}{\partial x_i \partial x_j} + \frac{\partial k_{ij}}{\partial x_i}\frac{\partial T}{\partial x_j} = \rho\, c_v \frac{\partial T}{\partial t}. \tag{38}$$

Simplifying the problem further to the case of orthogonal heat conduction, where $k_{ij} = 0$ for $i \neq j$ we have

$$\sum_{i=1}^{3}\left(k_{ii}\frac{\partial^2 T}{\partial x_i \partial x_i} + \frac{\partial k_{ii}}{\partial x_i}\frac{\partial T}{\partial x_i}\right) = \rho\, c_v \frac{\partial T}{\partial t}. \tag{39}$$

Let us focus on a 2-D heat transfer problem (Fig. 7), for the case when orthogonal heat conduction functions are known as $k_{11}(x_i)$ and $k_{22}(x_i)$.

We partition the domain in space using a mesh with uniform intervals: Δx_1 and Δx_2. In time the uniform interval will be denoted by $\Delta t^n = t^{n+1} - t^n$, where t^{n+1} and t^n are two consecutive time points. Let us denote the numerical approximation of the temperature function in the following way: $T(x_1^i, x_2^j, t^n) = \hat{T}_{ij}^n$. Then the derivatives in Eq. (39) can be expressed by finite differences. Using a forward difference at time t^n and a second order central difference for the space derivative at position (x_1^i, x_2^j) (FCTS method) we get from (39) the recurrence equation

$$[k_{11}]_{ij}\frac{\hat{T}_{i-1j}^n - 2\hat{T}_{ij}^n + \hat{T}_{i+1j}^n}{(\Delta x_1)^2} + [k_{22}]_{ij}\frac{\hat{T}_{ij-1}^n - 2\hat{T}_{ij}^n + \hat{T}_{ij+1}^n}{(\Delta x_2)^2}$$
$$+ \left[\frac{\partial k_{11}}{\partial x_1}\right]_{ij}\frac{\hat{T}_{i+1j}^n - \hat{T}_{i-1j}^n}{2\Delta x_1} + \left[\frac{\partial k_{22}}{\partial x_2}\right]_{ij}\frac{\hat{T}_{ij+1}^n - \hat{T}_{ij-1}^n}{2\Delta x_2} = [\rho\, c_v]_{ij}\frac{\hat{T}_{ij}^{n+1} - \hat{T}_{ij}^n}{\Delta t} \tag{40}$$

where $[\ ]_{ij}$ denotes quantities of functions included in brackets at position (x_1^i, x_2^j). By rearranging equation (40) one can get a solution for the problem i.e. the temperature in point (x_1^i, x_2^j) $\hat{T}_{ij}^{n+1}$ for the time, while t^{n+1}, when boundary conditions from S_1 to S_4 have been satisfied. The presented approach was applied to finding temperature distribution in circular plates with a composition showing plate thickness as in Fig. 3, Sadowski et al. (2007a).

4.   Transient thermal stress state

Having specified the temperature field $T(x_i, t)$ for the whole specimen or structural element, we can calculate the thermal stress. The governing differential equations of thermoelasticity are as follows:

1. The equations of motion

$$\bar{\sigma}_{ij,j} + \rho\, f_i = \rho\, \ddot{u}_i \quad \text{on} \quad V \tag{41}$$

where f_i are components of the body forces per unit mass and $\ddot{u}_i$ denotes the components of particle acceleration or equilibrium equation in quasi-static problems

$$\bar{\sigma}_{ij,j} + \rho\, f_i = 0 \quad \text{on} \quad V \tag{42}$$

2. Duhamel-Neuman constitutive equation

$$\bar{\sigma}_{ij} = \bar{C}_{ijkl}\, \bar{\varepsilon}_{kl} - \bar{\beta}_{ij}(T - T_0) \tag{43}$$

or in the reciprocal form

$$\bar{\varepsilon}_{ij} = \bar{S}_{ijkl}\, \bar{\sigma}_{kl} + \bar{\alpha}_{ij}(T - T_0) \tag{44}$$

where $\bar{\sigma}_{ij}$ and $\bar{\varepsilon}_{ij}$ are the averaged stress and strain symmetric tensors in CMCs; $\bar{C}_{ijkl}(x_m, T)$ and $\bar{S}_{ijkl}(x_m, T)$ are components of the averaged or equivalent stiffness and compliance elastic tensors of CMCs. In general the composite material is non-homogeneous, anisotropic and its properties can depend on space and temperature T. $\bar{\alpha}_{ij}(x_m, T)$ and $\bar{\beta}_{ij}(x_m, T)$ are second order symmetric tensors describing anisotropic and non-homogeneous thermal properties of the material. T_0 is the initial temperature of the thermal process.

3. The heat conduction equation in the most general form, taking into account thermo-mechanical coupling

$$\left(k_{ij} \frac{\partial T}{\partial x_j} \right)_{,i} = -Q + \rho\, c_v \frac{\partial T}{\partial t} + T_0 \bar{\beta}_{ij} \dot{\bar{\varepsilon}}_{ij}. \tag{45}$$

The simplest version without coupling effect describes Eq. (26).

The equation of motion can be expressed in relation to displacement vector components u_i by applying constitutive Eq. (43) and the definition of small strains

$$\frac{1}{2}\left[\bar{C}_{ijkl,j}\left(u_{k,l}+u_{l,k}\right)+\bar{C}_{ijkl}\left(u_{k,l}+u_{l,k}\right)_{,j}\right]+\rho\,f_{\mathrm{i}}=\rho\,\ddot{u}_{i}+\bar{\beta}_{ij}T_{,j}+\bar{\beta}_{ij,j}T$$
(46)

(45) and (46) constitute the set of governing equations for finding 4 unknown functions u_i and T. The transient stress state $\bar{\sigma}_{ij}(x_m,T)$ can be estimated from the Duhamel-Neuman constitutive Eq. (43).

The behaviour of thermally and mechanically isotropic and non-homogeneous elastic composite materials with perfect bonding at interphase boundaries is described by

- Young's modulus $\bar{E}(x_m,T)$

- Poisson's ratio $\bar{\nu}(x_m,T)$

- Thermal properties of the material: $\bar{\alpha}(x_m,T)$, $\bar{\beta}(x_m,T)$

The simplest approach to describing FGMs response is an elastic isotropic composite material model with only mechanical non-homogeneity. Its characteristics are denoted by $\bar{E}(x_m)$, $\bar{\nu}(x_m)$, $\bar{\alpha}(x_m)$, $\bar{\beta}(x_m)$.

This complex 3-D problem, can often be simplified to a 2-D case. In order to solve the temperature transient stress problem in FGMs, Airy's stress function is introduced

$$\bar{\sigma}_{ij}=-F_{,ij}+\delta_{ij}F_{,kk}$$
(47)

for $i,j,k=1,2$. Assuming that FGM material can be described by the simplest material model, and if the Beltrami-Michell equations are satisfied, one can get scalar equation with 2 unknowns: F and T

$$\frac{1-\bar{\nu}^2}{\bar{E}}\nabla^2\nabla^2F+2\left(\frac{1-\bar{\nu}^2}{\bar{E}}\right)_{,i}\left(\nabla^2F\right)_{,i}+\nabla^2\left(\frac{1-\bar{\nu}^2}{\bar{E}}\right)\nabla^2F$$
$$-\left(\frac{1+\bar{\nu}}{\bar{E}}\right)_{,ij}\left(\delta_{ij}\nabla^2F-F_{,ij}\right)+\nabla^2\left[(1+\bar{\nu})\,\bar{\alpha}(T-T_0)\right]=0 \qquad \text{on}\quad V$$
(48)

Taking into account (45) and (48), we can solve the problem and find the thermal stress $\bar{\sigma}_{ij}(x_m,T)$, e.g. Jin and Batra (1998) and Noda et al. (2000).

5.   Thermal residual stress due to technological cooling process

5.1   Analytical models

The thermal residual stress created during technological cooling of the FGMs from the processing temperature has a significant influence on the material behaviour. To illustrate this problem let us consider a beam or strip as shown

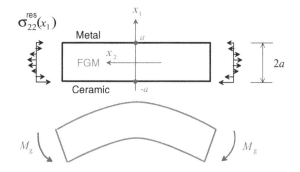

Figure 8.   Bending due to residual stress after cooling process

in Fig. 8, made of FGM composite. Two different types of gradation could be taken into account: continuous and discrete. With the assumption that x_1 is the symmetry axis, the analysed example simplifies to a1-D problem, i.e. all functions describing FGM strip response are related to the co-ordinate x_1. Let us introduce a dimensionless co-ordinate $\xi_1 = x_1/a$. Satisfying the stress free boundary conditions after a cooling process, one can get the thermal residual stress (e.g. Ravichandran, 1995; Sadowski and Neubrand, 2004)

$$\bar{\sigma}_{22}^{\text{res}}(\xi_1,\tilde{t}) = \Delta T^{\text{res}}\bar{E}(\xi_1,\tilde{t})\left[\bar{\alpha}(\xi_1,\tilde{t}) - \frac{A_1}{E_1} + \frac{\left(A_2 - \frac{A_1}{E_1}E_2\right)(\xi_1 E_1 - E_2)}{E_1 E_3 - E_2^2}\right]$$

(49)

where ΔT_{res} denotes the difference between the process temperature and room temperature, $\tilde{t} = (\kappa_0/a^2)t$ is the non-dimensional time, κ_0 is the reference thermal diffusivity at the ceramic surface of the specimen ($\xi_1 = -a$). Functions A_1 and A_2 depend on thermal and elastic properties of FGMs

$$A_1(\tilde{t}) = \int_{-1}^{1} \bar{\alpha}(\xi_1,\tilde{t})\bar{E}(\xi_1,\tilde{t})d\xi_1$$

(50)

$$A_2(\tilde{t}) = \int_{-1}^{1} \bar{\alpha}(\xi_1,\tilde{t})\bar{E}(\xi_1,\tilde{t})\xi_1 d\xi_1$$

(51)

Functions E_1, E_2, E_3 depend on elastic properties and are given by

$$E_1(\tilde{t}) = \int_{-1}^{1} \bar{E}(\xi_1,\tilde{t})d\xi_1$$

(52)

$$E_2(\tilde{t}) = \int\limits_{-1}^{1} \bar{E}(\xi_1, \tilde{t}) \xi_1 d\xi_1 \tag{53}$$

$$E_2(\tilde{t}) = \int\limits_{-1}^{1} \bar{E}(\xi_1, \tilde{t}) \xi_1^2 d\xi_1 \tag{54}$$

Calculations performed by Ravichandran (1995) for Al_2O_3/Ni type of FGM were done for: fully graded structure (Fig. 8), multilayered FGM system (like in Fig. 3), bi-layered material made of only Al_2O_3 and Ni, (structure shown in Fig. 5). The major conclusion is that a linear gradation of constituents gives the least residual stress after the cooling process. The character of the residual stresses in the ceramic region depends on the type of gradation profile. Application of multilayered FGM systems with many layers also leads to small residual stresses.

The second, more advanced model, was proposed by Freund (1993, 1996, 2000). It gives, in closed form, all elementary formulae for the stress distribution and curvature for a thin structural element consisting of several layers (e.g. Figs. 3 and 5) with arbitrary compositional gradation along the thickness, variation of temperature and epitaxial mismatch strain.

Let us consider a thin film, Fig. 5 and assume that it is subjected to the temperature difference $\Delta T = T - T_0$ and mismatch strain. All surfaces of the structure are free of applied traction. Then $\bar{\sigma}_{13} = \bar{\sigma}_{23} = \bar{\sigma}_{33} = 0$ and the remaining stress components depend on the non-dimensional coordinate $\xi_1 = x_1/a$, i.e. $\bar{\sigma}_{ij}(\xi_1, \tilde{t})$ for $i, j = 1,2$. The considered element remains in an equilibrium state, which means that net force, bending moment and twisting moment must vanish

$$\int\limits_{-1}^{1} [\bar{\sigma}_{ij}(\xi_1, \tilde{t})] d\xi_1 = 0, \qquad \int\limits_{-1}^{1} [\xi_1 \bar{\sigma}_{ij}(\xi_1, \tilde{t})] d\xi_1 = 0 \tag{55}$$

Assume that the response of CMCs or MCMs material is linearly elastic (43)

$$\bar{\sigma}_{ij}(\xi_1, \tilde{t}) = \bar{C}_{ijkl}(\xi_1, \tilde{t}) \bar{\varepsilon}_{kl}(\xi_1, \tilde{t}) - \bar{\beta}_{ij}(\xi_1, \tilde{t}) \Delta T(\xi_1, \tilde{t}) \tag{56}$$

and split the total strain $\bar{\varepsilon}_{ij}(\xi_1, \tilde{t})$ into compatible parts $\bar{\varepsilon}_{ij}^c(\xi_1, \tilde{t})$ and an incompatible one $\bar{\varepsilon}_{ij}^m(\xi_1, \tilde{t})$ (mismatch strain)

$$\bar{\varepsilon}_{ij}(\xi_1, \tilde{t}) = \bar{\varepsilon}_{ij}^c(\xi_1, \tilde{t}) + \bar{\varepsilon}_{ij}^m(\xi_1, \tilde{t}) \tag{57}$$

The compatibility conditions are satisfied for $\bar{\varepsilon}_{ij}^c(\xi_1, \tilde{t})$

$$\bar{\varepsilon}_{ij}^c(\xi_1, \tilde{t}) = -\xi_1 \hat{\kappa}_{ij}(\xi_1, \tilde{t}) + \bar{\varepsilon}_{ij}^r(\xi_1, \tilde{t}) \tag{58}$$

Here $\hat{\kappa}_{ij}$ is the curvature tensor, whereas $\bar{\varepsilon}^{\mathrm{r}}_{ij}(\xi_1, \tilde{t})$ is the strain tensor of reference layer $\xi_1 = 0$.

Introducing constitutive equation for CMC (56) and the strain tensor (57) to equilibrium Eq. (55) one can get the solution

$$\hat{\kappa}_{ij}(a, \tilde{t}) = \frac{I^{(1)} J^{(0)}_{ij} - I^{(0)} J^{(1)}_{ij}}{(I^{(1)})^2 - I^{(0)} I^{(1)}}, \tag{59}$$

$$\varepsilon^{\mathrm{r}}_{ij}(a, \tilde{t}) = \frac{I^{(2)} J^{(0)}_{ij} - I^{(1)} J^{(1)}_{ij}}{(I^{(1)})^2 - I^{(0)} I^{(1)}} \tag{60}$$

where

$$I^{(k)}(a, \tilde{t}) = \int_{-a}^{a} \xi_1^k \hat{E}(\xi_1, \tilde{t}) \mathrm{d}\xi_1 \tag{61}$$

and

$$J^{(k)}_{ij}(a, \tilde{t}) = \int_{-a}^{a} \xi_1^k \hat{E}(\xi_1, \tilde{t}) \left[\beta_{ij}(\xi_1^i \tilde{t}) \Delta T(\xi_1^i \tilde{t}) + \varepsilon^{\mathrm{m}}_{ij}(\xi_1^i \tilde{t}) \right] \mathrm{d}\xi_1 \tag{62}$$

In (61) and (62) the function $\hat{E}(\xi_1, \tilde{t})$ characterizes the elastic properties of CMCs, i.e. components of the stiffness tensor $\bar{C}_{ijkl}(\xi_1, \tilde{t})$; it depends on Young's modulus $\bar{E}(\xi_1, \tilde{t})$ and Poisson's ratio $\bar{\nu} = \mathrm{const.}$ describing the elastic response of CMC

$$\hat{E}(\xi_1, \tilde{t}) = \begin{cases} \bar{E}(\xi_1, \tilde{t})/(1 - \bar{\nu}^2) & \text{for} \quad i = j \\ \bar{E}(\xi_1, \tilde{t})/(1 - \bar{\nu}^2) & \text{for} \quad i \neq j \end{cases} \tag{63}$$

The formulas (59) and (60) allow for calculation of the curvature tensor and the strain of the reference plane, on condition that the material properties and mismatch strains are known.

Application of this theory was presented by Freund (1996, 2000) for a thin plate-like structure of square shape as in Fig. 5 or circular shape (Fig. 9). The analysis was limited to the existence of mismatch strains in the TBC structure. The presented results determined the limit of the linear and nonlinear range in

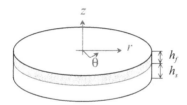

Figure 9. TBC structure of circular shape

the uniform relationship mismatch strain and of the curvature. Bifurcation of the deformation mode was also analysed.

Another analytical model, a so-called 3 parameter model, was proposed by Hsueh and Evans (1985), Hsueh (2002a,b), and Hsueh et al. (2003a,b). In this uniaxial bending theory, both lattice mismatch and thermal mismatch were analysed in layered and graded structural systems, e.g. Fig. 5. The three parameters, the uniform strain component, the position of bending axis and the curvature of the system, were calculated on the basis of the equilibrium conditions. The curvature and the stress distribution in Hsueh's and Freund's models are identical. Generalisation of the closed-form solution for elastic deformation of multilayers due to residual stresses and external bending was presented in Hsueh (2002b).

5.2   Numerical models

Residual stresses in FGMs can be calculated using an FEA approach.

Dao et al. (1997) estimated the values of residual stresses in a ceramic-FGM-metal three layer structure (Fig. 5) by using continuous and discrete micromechanic models. The influences of discrete microstructure on the residual stress distribution was estimated at grain size level with respect to material gradient and volume metal content. The plastic behaviour of metal grains was modeled at the single crystal level by applying crystal plasticity theory. The thermoelastic and thermoplastic deformations were taken into calculations. The results were compared with the continuous model which does not incorporate randomness and discreteness of the microstructure. Both models give a similar level of macroscopic stress. The discrete microstructural model predicts fairly high residual stress concentrations at the grain level.

Becker et al. (2000) assessed the thermoelastic response of FGM in various geometries: cylinder, thin film, FGM sphere and beam. The residual stresses were computed for a given temperature change ΔT by applying spatial derivatives of the Goodier potential, which were obtained from a weighted integration of the temperature field over the structure. The volume content of the ceramic phase V^C was introduced as a polynomial function of x_1 (e.g. Fig. 5), while the coefficient of thermal expansion was assumed to be a linear function of composition. The residual stresses were calculated using the computer algorithm FEAP (Zienkiewicz and Taylor, 1987). The numerical solutions were compared to analytical solutions.

A bi-layer laminated ceramic membrane subjected to bi-axial flexure was analysed by Atkinson and Selçuk (1999). The bi-layer membrane made of gadolinia-dopped ceria (180 μm thick) and yttria-stabilized zirconia (5 μm thick) contained significant residual stresses, which are consistent with elastic thermal expansion mismatch strains on cooling from the stress free state at 1200°C. The residual stresses were calculated using ABACUS code.

Schmauder et al. (2003) analysed the effect of residual stresses on the global mechanical properties of heterogeneous materials such as Al/SiC and Al/Al$_2$O$_3$ MMCs. It was found that residual stresses do have some impact on failure initiation in two-dimensional microstructures. However, irregular particle shapes are much more prone to fracture in a matrix than regular shapes. The calculation was done with MSC/PATRAN.

6. Basic fracture mechanics concepts in functionally graded materials (FGM)

Progressive cracking is a macroscopic manifestation of the net loss of inter-atomic bonds inside a brittle composite (CMC). This phenomenon is due to the nucleation of new, and growth of existing, microcracks. A microcrack can nucleate when the local nucleation strength of the material is exceeded by the local tensile or shear stress, i.e. if $\sigma_{ij}(x_i) \geq \sigma^{nucl}(x_i)$. Further crack growth is possible when the elastic energy release rate $\bar{G}$ (crack driving force) stored in the vicinity of its tip (described by co-ordinates x_i^b) exceeds the thermodynamic force G_R (crack resistance force) resisting its growth, i.e. when

$$\bar{G}[l, \sigma_{ij}(x_i^b), \Delta T, t] \geq G_R(l, x_i^b, \Delta T) \tag{64}$$

The energy release rate $\bar{G}$ depends on the elastic energy and potential energy of applied forces, and on the crack length l. In the general case it is related to the stress intensity factors (SIF) at the crack tip: K_I, K_{II}, K_{III}.

During the crack initiation and the subsequent progression of quasi-static crack advance, the total energy release rate must be equal to the energy needed for the fracture process, i.e.

$$\bar{G}[l, \sigma_{ij}(x_i^b), \Delta T, t] = G_R(l, x_i^b, \Delta T) \tag{65}$$

The equilibrium state of the crack is stable under fixed loading conditions, provided that the crack resistance increases more rapidly with increasing crack length than the crack extension force

$$\left.\frac{\partial \bar{G}}{\partial l}\frac{\partial l}{\partial t}\right|_{\substack{\sigma_{ij}=\text{const.}\\ \Delta T=\text{const.}}} + \left.\frac{\partial \bar{G}}{\partial t}\right|_{\substack{\sigma_{ij}=\text{const.}\\ \Delta T=\text{const.}}} < \left.\frac{\partial G_R}{\partial l}\frac{\partial l}{\partial t}\right|_{\substack{\sigma_{ij}=\text{const.}\\ \Delta T=\text{const.}}} \tag{66}$$

The limit of stable crack growth is reached when the critical condition

$$\left.\frac{\partial \bar{G}}{\partial l}\frac{\partial l}{\partial t}\right|_{\substack{\sigma_{ij}=\text{const.}\\ \Delta T=\text{const.}}} + \left.\frac{\partial \bar{G}}{\partial t}\right|_{\substack{\sigma_{ij}=\text{const.}\\ \Delta T=\text{const.}}} = \left.\frac{\partial G_R}{\partial l}\frac{\partial l}{\partial t}\right|_{\substack{\sigma_{ij}=\text{const.}\\ \Delta T=\text{const.}}} \cdot \tag{67}$$

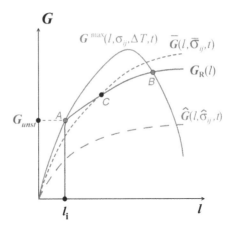

Figure 10. Possible manners of the crack behaviour

is reached. When the load is further increased, the equilibrium condition (65) is no longer fulfilled and the crack starts to propagate dynamically (becoming unstable).

Generally, a crack can grow only if at least one of the three SIF is positive. When all three stress intensity factors are negative a crack is dormant (does not grow) and almost always *passive* since it does not affect local deformations and effective parameters.

Figure 10 shows three possible behaviours of a straight crack under different types of loading $\hat{\sigma}_{ij}$, $\bar{\sigma}_{ij}$, σ_{ij} as a function of the characteristic crack length l. It also includes the crack resistance curve G_R for considered crack configuration. The function $\hat{G}(l, \hat{\sigma}_{ij}, t)$ describes the existence of "passive" cracks (which do not grow) or subcritical, stable crack development. The second function $\bar{G}(l, \bar{\sigma}_{ij}, t)$ describes the behaviour of the "active" crack, which becomes unstable beginning at point C. $G^{\max}(l, \sigma_{ij}, \Delta T, t)$ is an envelope of the maximum values of energy release rate, which take place during the thermal shock problems; the current state of stress σ_{ij} depends on actual time t. One can observe unstable crack growth between point A and B. At point B crack enlargement stops and reaches equilibrium length after thermal shock.

Microcracks have irregular shapes (e.g. intergranular, e.g. Sadowski, 1994), very seldom planar and embedded in CMCs, which cannot be treated as isotropic or homogeneous material. Therefore the exact determination of the thermodynamic force G is an extremely difficult task. Also the stresses σ_{ij} to which the microcracks are subjected could be influenced by the interaction with the adjacent microcracks and/or stress concentrations at the heterogeneities. On the other hand, the thermodynamic resisting force G_R (cohesive or fracture energy) depends on the architecture of the internal structure of the composite, i.e. distribution of energy barriers (grain boundaries, particle of fiber reinforcement, etc.). All these phenomena can strongly prevent or hinder the propagation of a microcrack, and are described at the macroscopic scale by R-curve behaviour, (Fig. 10). One can conclude that the functions $\bar{G}$ and G_R are random in relation to coordinates and internal structure of the composite. Therefore the progressive cracking process can evolve in heterogeneous materials by both crack nucleation and stable growth of cracks.

Because the progressive cracking process of the composites is intrinsic, its onset is related to critical values of the effective material properties ($\bar{C}_{ijkl}$ or $\bar{S}_{ijkl}$) and must be a function of the volume averages of stress $\bar{\sigma}_{ij}$ and strain $\bar{\varepsilon}_{ij}$; see constitutive Eq. (43) or (44). Therefore, the criteria of microcrack growth under thermal shock will be related to averaged characteristics describing the composite response.

According to Kawasaki and Watanabe (1993, 1997, 2002), thermal fracture behaviour of metal/ceramic FGMs always begins during the cooling process. The temperature of the first crack formation is defined as the critical surface temperature of the thermal shock. Due to high tensile stress at the ceramic (cooling) surface of a structural element, a system of edge cracks is generated (Fig. 11a). In relation to the internal structure, cracks can grow into FGMs in the direction of composite gradation, or can deflect due to encountering local energetic obstacles like strong metallic particles. For FGMs made of several graded layers or a single layer TBC, the edge cracks propagate straight to the first interface of the composite, and then can change direction of propagation, causing delamination of the interface between layers, and finally lead to spallation of the ceramic TBC (Fig. 11b).

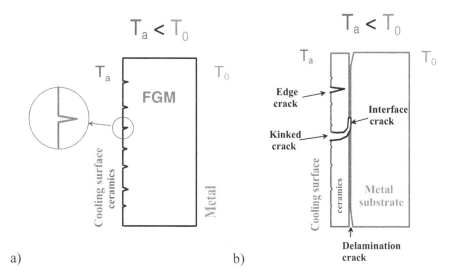

Figure 11. Cracks generated during thermal shock

Crack deflection starts when a small kink is created at the main crack tip. The kink of length l_k is inclined to the crack at an angle φ_k and its further growth is possible under a certain level of external load σ_{ij}. φ_k is estimated on the basis of the directional condition:

$$\left.\frac{\partial \bar{G}}{\partial \varphi}\right|_{\substack{l=\text{const}.\,\sigma_{ij}=\text{const}.\,\Delta T=\text{const}.\,t=\text{const}.}} = 0 \qquad (68)$$

The main crack propagates in a tangential direction from the kink tip when (68) and the failure condition

$$\bar{G}[\varphi_k, l_k, l, \sigma_{ij}(x_i^{\text{b}}), \Delta T, t] = G_R \qquad (69)$$

are satisfied.

The solution of Eqs. (68) and (69) requires the determination of the SIF at the kink tip: k_I and k_{II}. This is possible with numerical methods, e.g. Fujimoto and Noda (2001).

The limit of stable crack growth is reached when the critical condition is reached

$$\left.\frac{\partial \bar{G}}{\partial l_k}\frac{\partial l_k}{\partial t}\right|_{\substack{\sigma_{ij}=\text{const}.\\\Delta T=\text{const}.\\\varphi_k=\text{const}.\\l=\text{const}.}} + \left.\frac{\partial \bar{G}}{\partial t}\right|_{\substack{\sigma_{ij}=\text{const}.\\\Delta T=\text{const}.\\\varphi_k=\text{const}.\\l=\text{const}.}} = \left.\frac{\partial G_R}{\partial l_k}\frac{\partial l_k}{\partial t}\right|_{\substack{\sigma_{ij}=\text{const}.\\\Delta T=\text{const}.\\\varphi_k=\text{const}.\\l=\text{const}.}} \qquad (70)$$

The toughening mechanisms of ceramic matrix with metal particle reinforcement FGMs include

- Crack front bowling by interaction of the crack front and metallic particles

- Crack deflection by particulates ahead of the straight crack

- Bridging by ductile particles in the wake of the crack tip

- Residual stress due to mismatch between the thermal expansion coefficients of the ceramic matrix and metallic particles

Crack deflection is observed when the toughening mechanism in the ceramic-rich part of FGM; bridging toughening is important in metal-rich composition of FGM.

Let us focus on crack growth from ceramic-rich region into the metal-rich region of FGM in the direction of the material gradation. We will study R-curve behaviour of FGMs, particularly for mode I.

6.1 Rule of mixture to estimate the fracture toughness in FGMs

The simplest way to estimate critical energy release rate in FGMs is to apply a RoM (e.g. Jin and Batra, 1996a,b), similar to Eq. (4). However it can lead to overestimate the fracture toughness, because the fracture toughness of pure metal is much higher than that of particles distributed in the ceramic matrix. Using RoM, we can express the critical energy release rate for mode I crack propagation of CMCs by

$$\bar{G}_{Ic}(x_i^b, T) = V^C(x_i^b) G_{Ic}^C(T) + V^M(x_i^b) G_{Ic}^M(T) \tag{71}$$

Here $\bar{G}_{Ic}(x_i^b, T)$ is the critical energy release rate of the CMC for plane strain

$$\bar{G}_{Ic}(x_i^b, T) = \frac{1 - \bar{\nu}^2(x_i^b, T, t)}{\bar{E}(x_i^b, T, t)} [\bar{K}_{Ic}(x_i^b, T)]^2 \tag{72}$$

$G_{Ic}^C(x_i^b, T)$ is the critical energy release rate of the ceramic

$$G_{Ic}^C(T) = \frac{1 - [\nu^C(T, t)]^2}{E^C(T, t)} [K_{Ic}^C(T)]^2 \tag{73}$$

and $G_{Ic}^M(T)$ is the critical energy release rate of the metallic phase

$$G_{Ic}^M(T) = \frac{1 - [\nu^M(T, t)]^2}{E^M(T, t)} [K_{Ic}^M(T)]^2 \tag{74}$$

$\bar{K}_{Ic}(x_i^b, T)$, $K_{Ic}^C(T)$ and $K_{Ic}^M(T)$ are the fracture toughnesses of the FGM, ceramic and metal. Introducing (72)–(74) to (71) we find that the fracture toughness for the FGM is equal to

$$\frac{\bar{K}_{Ic}(x_i^b, T)}{K_{Ic}^C(T)} = \left\{ \frac{\bar{E}(x_i^b, T)}{1 - \bar{\nu}^2(x_i^b, T)} \left[\begin{array}{c} V^C(x_i^b) \frac{[1 - \nu^C(T)]^2}{E^C(T)} \\ + V^M(x_i^b) \frac{[1 - \nu^M(T)]^2}{E^M(T)} \left(\frac{K_{Ic}^M(T)}{K_{Ic}^C(T)} \right)^2 \end{array} \right] \right\} \tag{75}$$

The fracture toughness of the composite $\bar{K}_{Ic}(x_i^b, T)$ depends, in general, on the position x_i and current temperature T. The fracture toughness of a metal is much higher than the toughness resulting from distributed particles in ceramic materials (Krstic, 1983). Moreover, fracture toughness of a composite decreases with temperature increase, because the corresponding toughness of CMC components decrease, particularly in metallic phase. Bridging toughening is more important for metal rich CMCs.

As in (71), one can postulate the fracture toughness relation for mode II crack propagation in an FGM

$$\bar{G}_{IIc}(x_i^b, T) = V^C(x_i^b) G_{IIc}^C(T) + V^M(x_i^b) G_{IIc}^M(T) \tag{76}$$

Then it is possible to calculate the fracture toughness for crack propagation under mixed mode (mode I and II, e.g. Papadopoulos, 1993; Gross and Seelig, 2006) and for plane strain conditions

$$
\begin{aligned}
\bar{G}_{(I+II)c}(x_i^{\mathrm{b}}, T) &= \bar{G}_{Ic}(x_i^{\mathrm{b}}, T) + \bar{G}_{IIc}(x_i^{\mathrm{b}}, T) \\
&= \frac{1 - \bar{\nu}^2(x_i^{\mathrm{b}}, T)}{\bar{E}(x_i^{\mathrm{b}}, T)} \left\{ [\bar{K}_{Ic}(x_i^{\mathrm{b}}, T)]^2 + [\bar{K}_{IIc}(x_i^{\mathrm{b}}, T)]^2 \right\}
\end{aligned}
\tag{77}
$$

This equation is also useful for descriptions of kinked crack propagation.

6.2 Crack-bridging approach to assessing fracture toughness

After the edge cracks have initiated (Fig. 12), they will grow in the ceramics with metal particles which are plastically stretched behind the crack tip bridging the crack faces. Simplifying the problem of the further crack propagation, we assume that the FGM behaves elastically in front of the crack tip – metal particles do not deform plastically (e.g. Jin and Batra 1996a). Then a softening bridging law relates bridging stress $\sigma_{22}^{\mathrm{B}}(x_1, T)$ to the crack opening displacement $\delta_{22}^{B}(x_1, T)$ could be proposed

$$
\frac{\sigma_{22}^{\mathrm{B}}(x_1, T)}{\sigma_{22}^0} = \left(1 - \frac{\delta_{22}^{\mathrm{B}}(x_1, T)}{\delta_{22}^0} \right)^{\mathrm{nB}}
\tag{78}
$$

where σ_{22}^0 and δ_{22}^0 are the maximum bridging stress and the crack opening displacement, respectively: nB is the exponent describing the nonlinearity in the bridging law.

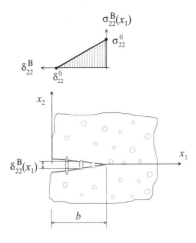

Figure 12.   Crack bridging zone

The bridging energy G^{B} is defined by

$$G^{\mathrm{B}} = \int\limits_0^{\delta_{22}^0} \sigma_{22}^{\mathrm{B}}(\delta_{22}^{\mathrm{B}})\mathrm{d}\delta_{22}^{\mathrm{B}} \tag{79}$$

Solving the integral equation for the crack problem, one can find the dislocation density function along the crack face, and hence the crack opening displacement function $\delta_{22}^B(x_1, T)$. The SIF at the crack tip consists of two parts: due to mechanical loading and crack bridging:

$$\bar{K}_I^{\mathrm{tip}} = \bar{K}_I^{\mathrm{mech}} + \bar{K}_I^{\mathrm{B}} \tag{80}$$

The initiated crack will grow in the ceramics when energy release rate is equal to the fracture toughness (65). According to (71) and (72), the critical value of the energy release rate will be calculated as

$$\bar{G}_{Ic}(x_1^{\mathrm{b}} = b, T) = V^{\mathrm{C}}(x_1^{\mathrm{b}} = b)\frac{1 - \bar{\nu}^2(x_1^{\mathrm{b}} = b, T)}{\bar{E}(x_1^{\mathrm{b}} = b, T)}[K_{Ic}^{\mathrm{C}}(x_1^{\mathrm{b}} = b, T)]^2 \tag{81}$$

and the crack growth criterion has the following form

$$\bar{K}_I^{\mathrm{tip}}(x_1^{\mathrm{b}} = b, T) = \left\{ V^{\mathrm{C}}(x_1^{\mathrm{b}} = b)\frac{1 - \bar{\nu}^2(x_1^{\mathrm{b}} = b, T)}{1 - [\nu^{\mathrm{C}}(T)]^2}\frac{E^{\mathrm{C}}(T)}{\bar{E}(x_1^{\mathrm{b}} = b, T)} \right\}^{0.5} K_{Ic}^{\mathrm{C}}(T) \tag{82}$$

Numerical calculations (Jin and Batra, 1996a) lead to the conclusion that the application of RoM, to assess the critical value of the energy release rate, overestimates the fracture toughness. However, further experimental investigations should be done. In particular the effect of the temperature influence should be experimentally tested.

7. Non-symmetric thermal shock in monolithic ceramic and FGM strip

Consider an infinitely long strip made of FGM (ceramic matrix Al_2O_3 and Al particles) (Sadowski and Neubrand, 2004). Both the mechanical and the thermal properties of the strip change gradually along the x_1 direction. Assume that the strip has initial temperature T_0. In non-symmetrical thermal shock the ceramic side of the strip ($x_1 = 0$) is cooled to the temperature T_a. The opposite side of the strip ($x_1 = h$) has the initial temperature T_0. Due to high concentration of the tensile thermal stress $\bar{\sigma}_{ij}^{\mathrm{T}}$ (Fig. 13b) an edge crack is initiated at the cooling ceramic surface of the strip (Fig. 14a). The generated crack will propagate in mode I as long as the thermal SIF exceeds the threshold value of the crack resistance of the FGM – $\bar{K}_{Ic}(x_i, T)$.

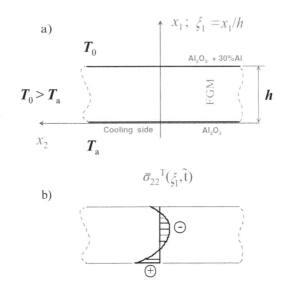

Figure 13. Strip under thermal shock (**a**) dimensions, (**b**) thermal stress at time t

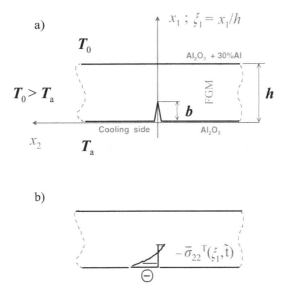

Figure 14. Edge crack (**a**) configuration and dimensions, (**b**) applied thermal stress at time t

The aim of this work is to estimate the equilibrium length $b(h, \Delta T, t)$ of the edge crack after thermal shock in FGM with continuous variation of the composition in relation to x_1 (e.g. $\bar{E}(x_1)$) (Fig. 14a).

7.1   Transient temperature distribution during thermal shock

Let us assume that there is no heat generation rate per unit volume Q. Then the heat conduction Eq. (25) simplifies to the one dimensional case:

$$\frac{d^2T}{d\xi_1^2} + \frac{1}{k}\left(\frac{dk}{d\xi_1}\frac{dT}{d\xi_1}\right) = \frac{1}{\kappa}\frac{dT}{d\tilde{t}} \tag{83}$$

Here the thermal diffusivity is equal to $\kappa = k/\rho c_v$ and $\tilde{t}$ is a non-dimensional time $\tilde{t} = (\kappa_0/h^2)t$, moreover $\kappa_0 = \kappa(\xi_1 = 0)$ and $\xi_1 = x_1/h$. Distribution of the temperature function $T(\xi_1, \tilde{t})$ was calculated for the following boundary and initial conditions $T(\xi_1 = 0, \tilde{t}) = T_a$ and $T(\xi_1 = 1, \tilde{t}) = T_0$.

The temperature field $T(\xi_1, \tilde{t})$ was obtained by numerical integration of (84) using the Runge-Kutta method for given thermal conductivity and thermal diffusivity functions of the form $k(\xi_1) = \sum k_i\xi_1^i$ and $\kappa(\xi_1) = \sum \kappa_i\xi_1^i$. In the present work the coefficients of these functions were estimated from experimental data.

7.2   Thermal and residual stresses

Let us assume that during the whole thermal shock process the strains in the CMC composite are small, i.e.

$$\varepsilon_{ij} = \frac{1}{2}[u_{i,j} + u_{j,i}] \tag{84}$$

Hooke's law (43) for isotropic elastic material subjected to thermal loading has the following form

$$\bar{\sigma}_{ij}(\xi_1, T, \tilde{t}) = \frac{\bar{E}}{1+\bar{\nu}}\left[\frac{\bar{\nu}}{1-2\bar{\nu}}\bar{\varepsilon}_{kk}\delta_{ij} + \bar{\varepsilon}_{ij}\right] - \bar{\alpha}\Delta T\frac{\bar{E}}{1-\bar{\nu}}\delta_{ij} \tag{85}$$

The material properties do not depend on temperature, e.g. $\bar{E}(\xi_1, \tilde{t})$, $\bar{\nu}(\xi_1, \tilde{t})$ and $\bar{\alpha}(\xi_1, \tilde{t})$.

Generated during thermal shock, thermal and residual stresses should satisfy the equilibrium Eq. (42). Excluding the body forces, we have

$$\frac{d^2}{d\xi_1^2}\left(\frac{1-\bar{\nu}^2}{\bar{E}}\bar{\sigma}_{22}^T\right) + \frac{d^2}{d\xi_1^2}[(1+\bar{\nu})\bar{\alpha}T] = 0 \tag{86}$$

Equation (86) was solved analytically for power law expansions (in relation to ξ_1) of the thermomechanical material properties and the stress free boundary conditions, i.e. no additional mechanical loading on the strip boundary. The particular form of the thermal stress function is as follows:

$$
\begin{aligned}
\bar{\sigma}_{22}^{\mathrm{T}}\left(\xi_1, T, \tilde{t}\right) = &-\frac{\bar{E}}{E^{\mathrm{C}}}\frac{\bar{\alpha}}{\alpha^{\mathrm{C}}}\frac{T}{\Delta T} \\
&+\frac{1}{1-(\bar{\nu})^2}\frac{\bar{E}}{E^{\mathrm{C}}A_0}
\left[
\begin{array}{l}
\{\xi_1 h A_{22} - A_{21}\} \int\limits_{-0}^{1} \bar{E}\frac{\bar{\alpha}}{\alpha_0}\frac{T}{\Delta T}h\mathrm{d}\xi_1 \\
+\{A_{11} - \xi_1 h A_{21}\} \int\limits_{-0}^{1} \xi_1 \bar{E}\frac{\bar{\alpha}}{\alpha_0}\frac{T}{\Delta T}h^2\mathrm{d}\xi_1
\end{array}
\right]
\end{aligned}
\tag{87}
$$

The functions $A_{ij}(\tilde{t})$ and $A_0(\tilde{t})$ depend on mechanical properties (Sadowski and Neubrand, 2004). The values $E^{\mathrm{C}} = \bar{E}(\xi_1 = 0, \tilde{t})$ and $\alpha^{\mathrm{C}} = \bar{\alpha}(\xi_1 = 0, \tilde{t})$ correspond to the ceramic edge, which is subjected to cooling. The exemplary stress distribution was presented in Fig. 13b.

The residual stresses were calculated according to (49).

7.3 Thermal stress intensity factor

Let us consider an FGM with a small gradation of the properties, i.e. the volume content of Al particles is up to $V^{\mathrm{M}} = 0.3$ in the Al_2O_3 matrix. Because the length of the crack after thermal shock is short in comparison to the specimen thickness, one can simplify the problem of crack propagation by assuming a constant Young's modulus inside the strip $E^{\mathrm{C}} = \bar{E}(\xi_1 = 0, \tilde{t})$. Then the thermal stress intensity factor in the FGM strip, Fig. 14b, can be found in a similar way as described in Jin and Batra (1996b), Erdogan and Wu (1996), and Noda (1999). This solution is made possible using an integral equation for the given tractions $p_{22}(\xi_1, \tilde{t}) = -\bar{\sigma}_{22}^{\mathrm{T}}(\xi_1, \tilde{t})$ on the crack boundary for

$$
\int\limits_{0}^{b/h} \frac{V_2(\xi_1, r, \tilde{t})}{(r - \xi_1)^2}\mathrm{d}r + \int\limits_{0}^{b/h} V_2(\xi_1, r, \tilde{t})K(\xi_1, r)\mathrm{d}r = -\left(\frac{4\pi}{E^{\mathrm{C}}}\right)p_{22}(\xi_1, \tilde{t})
\tag{88}
$$

Here $V_2(\xi_1, r, \tilde{t})$ is the crack opening displacement for the crack edges, and $K(\xi_1, r)$ is the integral kernel, e.g. Kaya and Erdogan (1987) and Sadowski (2005). The solution of (88) gives the thermal stress intensity factor function, which can be expressed in the dimensionless form

$$
\tilde{K}_I^{\mathrm{T}}(b, h, \tilde{t}) = \frac{1 - \nu^{\mathrm{C}}}{E^{\mathrm{C}}\alpha^{\mathrm{C}}\Delta T\sqrt{\pi h}}K_I(b, \tilde{t})
\tag{89}
$$

7.4 Numerical example

The Al_2O_3/Al FGM material was chosen for calculations because it is one of the rare FGMs for which all thermomechanical properties including residual stresses and crack growth resistance have been studied experimentally in detail (Chung et al., 2001; Neubrand et al., 2002) . For the purpose of this work, the properties of the analysed Al$_2$O$_3$/Al FGM were expressed as a function of the volume content (V^M) for the metallic phase Al. Young's modulus at room temperature (in GPa) was described by the following polynomial

$$\bar{E}(V^{Al}) = -1482(V^{Al})^3 + 1973(V^{Al})^2 - 1096(V^{Al}) + 398 \qquad (90)$$

whereas the thermal expansion coefficient $\bar{a}$ (in 10^{-6}/K) of the CMC was described by

$$\bar{a}(V^{Al}) = 281(V^{Al})^3 - 103(V^{Al})^2 - 15(V^{Al}) + 7.71 \qquad (91)$$

The appropriate function for thermal conductivity k (in W/mK) is given by

$$\bar{k}(V^{Al}) = 37.71 + 363(V^{Al})^{1.45} \exp(-1.5V^{Al}) \qquad (92)$$

and the thermal diffusivity κ (in cm^2/s) is

$$\bar{\kappa}(V^{Al}) = 0.109 + 1.844(V^{Al})^{1.5} \exp(-2.5V^{Al}) \qquad (93)$$

In the example the strip had a width of $h = 10$ mm.

The composition gradient along the ξ_1 – direction is described by the power function

$$V^{Al}(n, \xi_1) = V_0^{Al} + V_n^{Al}\xi_1^n \qquad (94)$$

Here, V_0^{Al}, V_n^{Al} and n are material parameters. By introducing (94) to (90)–(93) and varying the material parameters, we can investigate the thermal shock response of different FGMs. In the numerical example, $V_0^{Al} = 0.03$ and $V_n^{Al} = 0.3$. For the considered composition gradient, the FGM material behaves macroscopically brittle, and thus plasticity can be neglected. For $n = 1$, we have a linear composition gradient of the Al in Al$_2$O$_3$. For $n < 1$, the metal content increases quickly in FGM and we have the case of a "metal rich material"; for $n > 1$ the metal content increases slowly, and we have a "ceramic rich material".

Figure 15 presents the temperature distribution $T(\xi_1, \tilde{t})$ during thermal shock along strip thickness for data given by (92) and (93).

Thermal stress was calculated according to (87), using (89) and (91) for estimating the temperature function $T(\xi_1, \tilde{t})$. In the numerical examples, two thermal shock cases were considered for $\Delta T = 600$ K and $\Delta T = 300$ K. The cooling temperature was equal to $T_a = 293$ K.

For the calculation of the residual stresses, it was assumed that the composite was stress free at 893 K (at this temperature the aluminium in the composite is still very soft and cannot exert high stresses on the ceramic backbone of the composite, irrespective of volume content). The residual stresses during cooling to room temperature were calculated from the thermal expansion coefficients and elastic modulus data. It has been shown, for specimens of a

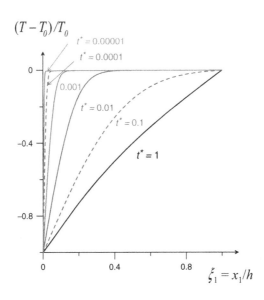

Figure 15. Temperature distribution during thermal shock along strip thickness

different geometry, that the stresses calculated with these assumptions are in reasonable agreement with experimental data (Neubrand et al., 2002).

The dimensionless thermal stress intensity factor $\tilde{K}_I^T$ depends on time, crack length and temperature. For a given crack length, the stress intensity factor will reach a maximum at a certain time $\tilde{t}$ which will increase with crack length. The envelope of the maximum values of the stress intensity factor $\tilde{K}_I^T$ for the "ceramic rich material" is plotted in Fig. 16 as a function of crack length ξ_1^b. Figure 16 also contains the R-curves for Al_2O_3/Al. The point where the curve for the maximum stress intensity factor intersects the R-curves represents the equilibrium crack length after thermal shock.

The envelope of the maximum values of the stress intensity factor $\tilde{K}_I^T$, the linear volume content of Al is plotted in Fig. 17 as a function of crack length ξ_1^b. The appropriate envelope for the "metal rich material" is plotted in Fig. 18. The point where the curve for the maximum stress intensity factor intersects the R-curves represents the equilibrium crack length after thermal shock.

Figure 19 represents the equilibrium crack length for the three FGM compositions, i.e. for values of $n = 1/3; 1; 3$. The crack lengths are much shorter in the graded Al_2O_3/Al composite than in a homogenous composite with a volume content of Al which corresponds to the cooling surface of the FGM ($\xi_1 = 0$).

Residual stresses in the graded composite are typically compressive near the cooling surface and lead to smaller values for the stress intensity factors and the equilibrium crack length (for longer cracks the stress intensity factors even

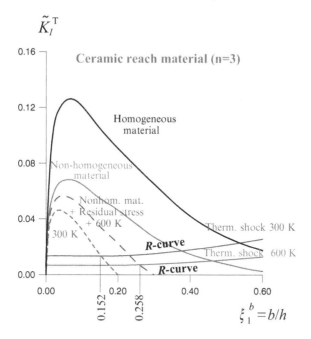

Figure 16. Non-dimensional thermal stress intensity factor $\tilde{K}_I^T$ for ceramic rich material

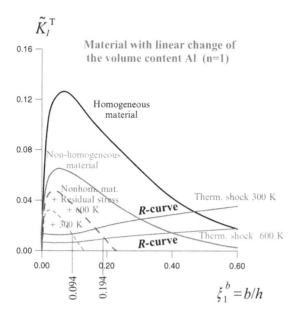

Figure 17. Dimensionless thermal stress intensity factor $\tilde{K}_I^T$ for $n = 1$

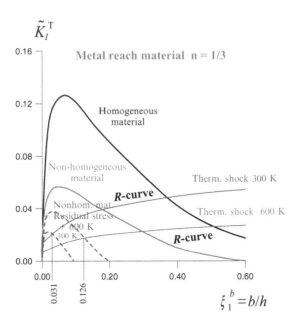

Figure 18. Dimensionless thermal stress intensity factor $\tilde{K}_I^T$ for "metal rich material"

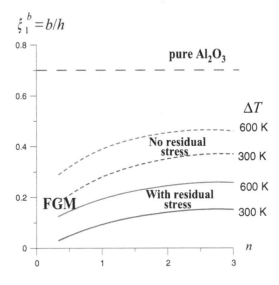

Figure 19. Edge crack (**a**) configuration and dimensions, (**b**) applied thermal stress at time t

become negative indicating very efficient crack arrest). The effect of residual stresses is strongest for "metal rich material" ($n = 1/3$) where the equilibrium crack length is reduced by about 60% compared to a hypothetical stress free material. The "metal rich" composite with $n = 1/3$ also shows the shortest equilibrium crack length. The short crack length is caused not only by the residual stresses, but also by the crack growth resistance of the material, which increases quickly due to its metal-rich composition profile. Close examination of Figs. 16–18 reveals that the graded material, with $n = 1/3$, would show the lowest stress intensity factor under thermal shock, even if residual stresses were absent. Figure 19 summarises these results both for thermal shock $\Delta T = 300\,\text{K}$ and $\Delta T = 600\,\text{K}$. As expected, the crack lengths after thermal shock by $300\,\text{K}$ are much less than those for $600\,\text{K}$.

The analysis confirms earlier calculations given by Noda (1999), which also predict that thermal shock stress intensity factors can be reduced by gradients. This contribution points out that significantly shorter equilibrium crack lengths can be expected if the rising crack growth (R-curve) of an FGM is combined with a controlled residual stress profile. Residual stress and the rising crack growth resistance should thus be included in any analysis of thermal shock resistance of functionally graded materials.

8. Two-dimensional thermal shock problem in layered circular plates

Consider a laminated circular plate made of five ceramic layers and diameters as shown in Fig. 20, and describe the FGM behaviour under non-symmetric thermal shock. The problem was investigated experimentally and analysed theoretically by Sadowski et al. (2007a, 2008b).

8.1 Samples preparation and experimental procedure

Plates with a 30 mm diameter and thickness 2.5 mm were made of monolithic Al_2O_3 (which is the basic material for FGM) and FGM. The specimen consists of five ceramic layers (each of them 0.5mm thick), (Fig. 20): a pure Al_2O_3 layer, and composite layers made of Al_2O_3 matrix and 5, 10, 15, 20 wt% content of ZrO_2. The CMC layers were fabricated from mixtures of commercially available oxides: alumina ZD Skawina (Poland) and 3 mol% Y_2O_3 stabilized zirconia Zhongshun, China. Each layer of the FGM samples were created in the appropriate raw volume in order to be 0.5 mm thickness after pressing (120 MPa) and final sintering at 1973 K for 1.5 h in air.

All experiments (Fig. 21) were performed using the following steps:

1. Introduction of the five initial cracks of the Vickers indenter in the upper central alumina side part of the plates. The applied load range was from 49 to 200 N

2. Measurement of the initial crack length c_o

3. Heating of the samples in a furnace up to the initial temperature $T_0 = 973\ \text{K}$

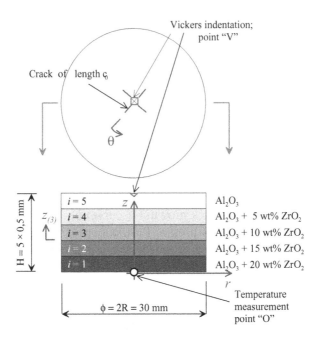

Figure 20. FGM circular plate specimen used for non-symmetric thermal shock testing

4. Sudden cooling of the specimens on the upper side using a high-velocity nitrogen jet (about 17 l/min) at room temperature. The gas was channelled onto the disk centre for about 20 s using a metal tube (inner diameter $D = 4.5$ mm) placed 3 mm above the test surface perpendicularly (with Vickers cracks)

5. Recording the temperature variation during the gas impinging of the lower sample surface in the central point, and concurrently at a peripheral one with the aid of PtRh/Pt thermocouples

6. Measurement of the final crack length c by an optical microscope

8.2   Theoretical formulation

8.2.1   Thermal and mechanical properties of monolithic and FGM material

For the purpose of heat transfer modelling, the thermal and mechanical properties of the material used for specimen preparation were evaluated according to the linear RoM, Eq. (1).

Data for the monolithic Al_2O_3 material are based on (Munro, 2001; Sadowski et al., 2006a) and the thermomechanical properties of pure zirconia

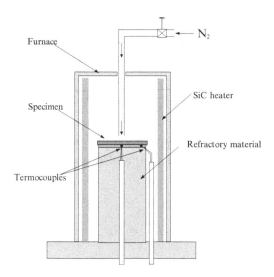

Figure 21.   Scheme of the thermal shock stand

were taken from (Stevens, 1986; Sadowski et al., 2006a). The densities have been estimated as $\rho^{Al_2O_3} = 3891$ kg/m^3 for Al_2O_3 and $\rho^{ZrO_2} = 6100$ kg/m^3 for ZrO_2, respectively.

The properties of the FGM material have been evaluated according to RoM, (1), i.e. using the zirconia volumetric fraction V^{ZrO_2} as the weight. For example, we use the following formula for calculating thermal conductivity of a single, ith layer of the FGM:

$$^{(i)}\bar{k} = {}^{(i)}k^{Al_2O_3/ZrO_2} = k^{Al_2O_3}\left(1 - V_{(i)}^{ZrO_2}\right) + k^{ZrO_2}V_{(i)}^{ZrO_2} \tag{95}$$

Additionally, we assume that every ith layer is thermally and mechanically isotropic. For example, it means that there is no difference between thermal conductivity in the r and z directions, $^{(i)}k_r^{Al_2O_3/ZrO_2} = {}^{(i)}k_z^{Al_2O_3/ZrO_2}$. It is necessary to point out that changes of $^{(i)}k_z^{Al_2O_3/ZrO_2}$ in the z direction are small because of the small variations of zirconia content within consecutive material layers (by 5% of weight fraction).

8.2.2   Heat conduction problem in FGM circular plate specimens

Let $b_{(i)}$ represent the thickness of the ith layer (Fig. 20), and, $z_{(i)}$ represent the local coordinate system of the ith layer, the origin of which is taken at the base of the ith layer. We assume that the specimen is initially at T_0 temperature and is suddenly cooled from the upper surface by using a high-velocity nitrogen jet placed at the central point of the plate. Thus the heat conduction analysis is axisymmetrical and is treated as a 2-D problem related to (r, z). The relative heat transfer coefficient at the top surface of the specimen is expressed by the function $h(r, z = H)$. Then, the transient heat conduction equation for the ith layer is given as

$$\bar{\rho}_{(i)}\bar{c}_{v(i)}\frac{\partial T_{(i)}}{\partial t} = {}^{(i)}\bar{k}_{\mathrm{r}}\left(\frac{1}{r}\frac{\partial T_{(i)}}{\partial r} + \frac{\partial^2 T_{(i)}}{\partial r^2}\right) + {}^{(i)}\bar{k}_{\mathrm{z}}\frac{\partial^2 T_{(i)}}{\partial z_{(i)}^2} + \frac{\partial^{(i)}\bar{k}_{\mathrm{z}}}{\partial T_{(i)}}\left(\frac{\partial T_{(i)}}{\partial z_{(i)}}\right)^2 \tag{96}$$

for $i = 1,\ldots,5$. Here $\bar{\rho}_{(i)}$, $\bar{c}_{v(i)}$ denote density of the ith layer and its specific heat, respectively. The solution of the problem should satisfy the following conditions:

- Initial boundary conditions

$$t = 0 \qquad T_{(i)} = T_0 \quad i = 1,\ldots,5 \tag{97}$$

- Thermal boundary conditions on the specimen surface

$$\begin{Bmatrix} z = 0, \\ z_{(1)} = 0 \end{Bmatrix} \qquad \frac{\partial T_{(1)}}{\partial z_{(1)}} = 0 \tag{98}$$

$$r = R \qquad \frac{\partial T_{(i)}}{\partial r} = 0 \quad \text{for} \quad i = 1,\ldots,5 \tag{99}$$

$$\begin{Bmatrix} z = H, \\ z_{(5)} = b_{(5)} \end{Bmatrix} \qquad \frac{\partial T_{(5)}}{\partial z_{(5)}} + h\left(T_{(5)} - T_0\right) = 0 \tag{100}$$

$$r = 0 \qquad \frac{\partial T_{(i)}}{\partial r} = 0 \quad \text{for} \quad i = 1,\ldots,5 \tag{101}$$

- Interfacial boundary conditions between adjacent layers for $i = 1,\ldots,4$

$$\begin{Bmatrix} z_{(i)} = b_{(i)}, \\ z_{(i+1)} = 0 \end{Bmatrix} \qquad T_{(i)} = T_{(i+1)} \tag{102}$$

$$\begin{Bmatrix} z_{(i)} = b_{(i)}, \\ z_{(i+1)} = 0 \end{Bmatrix} \qquad {}^{(i)}\bar{k}_{\mathrm{z}}\frac{\partial T_{(i)}}{\partial z_{(i)}} = {}^{(i+1)}\bar{k}_{\mathrm{z}}\frac{\partial T_{(i+1)}}{\partial z_{(i+1)}} \tag{103}$$

The temperature function for the whole plate is

$$T(r, z) = \sum_{i=1}^{5} T_{(i)}(r, z_{(i)}) \tag{104}$$

The solution of the heat conduction Eq. (96) including the initial and boundary conditions (97)–(103) was done by the application of FD approach (Sadowski et al., 2007a; Sadowski and Nakonieczny, 2008) or FEA approach (Sadowski et al., 2008a,b). It was additionally assumed that FGM layers are thermally isotropic, i.e. ${}^{(i)}\bar{k}_{\mathrm{r}} = {}^{(i)}\bar{k}_{\mathrm{z}}$.

8.2.3   Determination of the thermal stress

Let us derive the basic equations that govern the thermoelastic behaviour of the plate. Since circular plate is thin, we can assume that the plane perpendicular to the neutral plane ($z = H/2$) before deformation remains plane, and perpendicular to it after deformation. Moreover, the axial stress $\bar{\sigma}_{zz}$ is small compared with other stress components. Furthermore, $\bar{\sigma}_{r\theta}$, $\bar{\sigma}_{z\theta}$, $\bar{\varepsilon}_{r\theta}$ and $\bar{\varepsilon}_{z\theta}$ are equal to zero - axisymmetrical plate deformation. With these assumptions, in-plane strain components in each ith layer are given by

$$\bar{\varepsilon}_{rr}^{(i)}(r,z) = \frac{\partial u_r^{(i)}}{\partial r} = \bar{\varepsilon}_{rr0}^{(i)} - z\frac{\partial^2 u_z^{(i)}}{\partial r^2} \tag{105}$$

$$\bar{\varepsilon}_{\theta\theta}^{(i)}(r,z) = \frac{u_r^{(i)}}{r} = \bar{\varepsilon}_{\theta\theta 0}^{(i)} - z\frac{\partial u_z^{(i)}}{\partial r} \tag{106}$$

$$\bar{\varepsilon}_{zr}^{(i)}(r,z) = \frac{1}{2}\left(\frac{\partial u_z^{(i)}}{\partial r} + \frac{\partial u_r^{(i)}}{\partial z}\right) \tag{107}$$

$$\bar{\varepsilon}_{zz}^{(i)}(r,z) = \frac{\partial u_z^{(i)}}{\partial z} \tag{108}$$

where $\bar{\varepsilon}_{rr0}^{(i)}$ and $\bar{\varepsilon}_{\theta\theta 0}^{(i)}$ denote in-plane strain components on the neutral plane. $u_z^{(i)}$ is a deflection of ith layer. The thermoelastic stress–strain equations have the forms

$$\bar{\varepsilon}_{rr}^{(i)}(r,z) = \frac{1}{\bar{E}^{(i)}}\left[\bar{\sigma}_{rr}^{(i)} - \bar{\nu}^{(i)}\bar{\sigma}_{\theta\theta}^{(i)}\right] + \bar{\alpha}^{(i)}(T_{(i)} - T_0) \tag{109}$$

$$\bar{\varepsilon}_{\theta\theta}^{(i)}(r,z) = \frac{1}{\bar{E}^{(i)}}\left[\bar{\sigma}_{\theta\theta}^{(i)} - \bar{\nu}^{(i)}\bar{\sigma}_{rr}^{(i)}\right] + \bar{\alpha}^{(i)}(T_{(i)} - T_0) \tag{110}$$

$$\bar{\varepsilon}_{zz}^{(i)}(r,z) = -\frac{\bar{\nu}^{(i)}}{\bar{E}^{(i)}}\left[\bar{\sigma}_{rr}^{(i)} + \bar{\sigma}_{\theta\theta}^{(i)}\right] + \bar{\alpha}^{(i)}(T_{(i)} - T_0) \tag{111}$$

$$\bar{\varepsilon}_{zr}^{(i)}(r,z) = \frac{1+\bar{\nu}^{(i)}}{\bar{E}^{(i)}}\bar{\sigma}_{zr}^{(i)} \tag{112}$$

The equilibrium equations for axisymmetric thermal shock problem are

$$\frac{\partial\bar{\sigma}_{rr}^{(i)}}{\partial r} + \frac{\partial\bar{\sigma}_{zr}^{(i)}}{\partial z} + \frac{\bar{\sigma}_{rr}^{(i)} + \bar{\sigma}_{\theta\theta}^{(i)}}{r} = 0 \tag{113}$$

$$\frac{\partial\bar{\sigma}_{rz}^{(i)}}{\partial r} + \frac{\bar{\sigma}_{rz}^{(i)}}{r} = 0 \tag{114}$$

By substituting (107)–(112) into (113) and (114), after straightforward derivation, one can get a set of partial differential equations:

$$
\frac{\partial^2 u_r^{(i)}}{\partial r^2} + \frac{\partial u_r^{(i)}}{r \partial r} - \frac{u_r^{(i)}}{r^2} + \frac{(1 - \bar{\nu}^{(i)})}{2} \left(\frac{\partial^2 u_r^{(i)}}{\partial z^2} + \frac{\partial^2 u_z^{(i)}}{\partial r \partial z} \right)
$$

$$
+ \frac{[1 - (\bar{\nu}^{(i)})^2]^2}{\bar{E}^{(i)}} \frac{\bar{\alpha}^{(i)}}{r} (T_{(i)} - T_0) \tag{115}
$$

$$
- \frac{\bar{\nu}^{(i)}(1 + \bar{\nu}^{(i)})[1 - (\bar{\nu}^{(i)})^2]}{\bar{E}^{(i)}} \bar{\alpha}^{(i)} \frac{\partial T_{(i)}}{\partial r} = 0
$$

$$
\frac{\partial^2 u_z^{(i)}}{\partial r^2} + \frac{\partial^2 u_r^{(i)}}{\partial r \partial z} + \frac{1}{r} \left(\frac{\partial u_z^{(i)}}{\partial r} + \frac{\partial u_r^{(i)}}{\partial z} \right) = 0 \tag{116}
$$

Equations (115) and (116) yield the displacement functions: $u_r^{(i)}$ and $u_z^{(i)}$ for each layer, i.e. $i = 1, \ldots, 5$.

If the lower and upper surfaces are traction-free and the interfaces of each layer are perfectly bonded, then:

- The boundary conditions of the lower and upper surfaces are the following:

$$
\left\{ \begin{array}{l} z = 0, \\ z_{(1)} = 0 \end{array} \right\} \qquad \bar{\sigma}_{zz}^{(1)} = 0; \quad \bar{\sigma}_{zr}^{(1)} = 0 \tag{117}
$$

$$
\left\{ \begin{array}{l} z = H, \\ z_{(5)} = b_{(5)} \end{array} \right\} \qquad \bar{\sigma}_{zz}^{(1)} = 0; \quad \bar{\sigma}_{zr}^{(1)} = 0 \tag{118}
$$

- The conditions of continuity at the interfaces can be represented as

$$
\left\{ \begin{array}{l} z_{(i)} = b_{(i)}, \\ z_{(i+1)} = 0 \end{array} \right\} \qquad \bar{\sigma}_{zz}^{(i)} = \bar{\sigma}_{zz}^{(i+1)}; \quad \bar{\sigma}_{zr}^{(i)} = \bar{\sigma}_{zr}^{(i+1)} \tag{119}
$$

$$
\left\{ \begin{array}{l} z_{(i)} = b_{(i)}, \\ z_{(i+1)} = 0 \end{array} \right\} \qquad u_r^{(i)} = u_r^{(i+1)}; \quad u_z^{(i)} = u_z^{(i+1)} \tag{120}
$$

For a simply supported plate, the mechanical boundary conditions are given by

$$
r = R \qquad \sigma_{rr}^{(i)} = 0; \quad u_z^{(i)} = 0 \quad \text{for} \quad i = 1, \ldots, 5 \tag{121}
$$

The solution of this problem can be performed by the finite element method. For this paper the problem was calculated by LS-DYNA commercial code (Sadowski et al., 2008b).

8.3   Numerical example

The numerical solution of the heat conduction Eq. (96) by FD or FEA leads to the specification of the temperature function (104). The most important points lie along the symmetry axis (Fig. 22). The central part of the specimens are subjected to the highest temperature difference ΔT and the maximum of the thermal shock. Figures 22 and 23 represent distributions of the temperature during the thermal shock at the central point "O" of the bottom specimen side. The results were obtained by using a model of convective heat transfer proposed by Sadowski et al. (2007a), and agree with previous experiments.

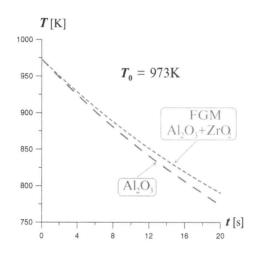

Figure 22.   Temperature distribution in relation to time for point "O" (Fig. 20)

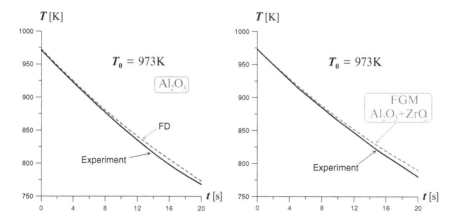

Figure 23.   Temperature distribution in relation to time for point "O" – comparison to experiments

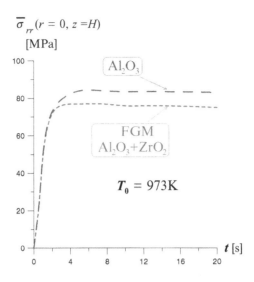

Figure 24.   Stress distribution in relation to time for central point at the upper surface of the specimen

In order to analyse crack propagation during thermal shock, several Vickers indentations were introduced to the top surface of the Al_2O_3 and FGM samples. The initial lengths of cracks c_0 were measured before the heating and sudden cooling processes. According to Trancert and Osterstock (1997) and Sadowski et al. (2007b) the relation between the fracture toughness to the indentation load P is

$$K_{Ic} = \chi P c_0^{-3/2} \tag{122}$$

where χ is a factor dependant on the elasto-plastic properties of the material and on the indenter geometry.

The thermal shock cooling process will cause high values of tensile stress at the top surface of the specimen $\bar{\sigma}_{rr}(r = 0, z = H)$, (Fig. 24). The tensile stress influences the crack propagation process. The lengths of the Vickers cracks extend to the dimension c up to the mechanical equilibrium point, which is expressed by

$$K_{Ic} = \chi P c^{-3/2} + \bar{\sigma}_{rr}(r = 0, z = H)\sqrt{\pi \Omega c} \tag{123}$$

The second term represents the stress intensity field of a semi-circular crack; $\Omega = 4/\pi^2$. Rewriting (32) and studying the derivative $d\sigma/dc$ allows us to predict and verify the existence of a stage of stable crack growth before the catastrophic failure of the indented sample occurs. This means that in a given range of post-indentation applied loading, the cracks will extend with increasingly applied stress, and cease to increase as soon as it diminishes. This is representative of transient stress. Therefore, the relative crack growth during quenching, c/c_0, can be used as an indicator of thermal shock resistance; one can also evaluate the maximum value of the thermal transient tensile stress σth. The stress σth can be calculated from (122) and (123):

$$\sigma_{\text{th}} = \frac{\chi P \left(c_0^{-3/2} - c^{-3/2}\right)}{(\pi \Omega c)^{1/2}} \tag{124}$$

In order to assess thermal shock, Trancert and Osterstock (1997) proposed using a parameter R_m,:

$$R_m = \left(\frac{K_{Ic}}{\sigma th}\right)^2 \tag{125}$$

The parameter R_m has the dimension of length, and is proportional to the maximum surface flaw that can be withstood by the material for the given thermal shock conditions:

$$R_m = \frac{\pi \Omega c}{[1 - (c/c_0)^{-3/2}]^2} \tag{126}$$

One can interpret the parameter R_m as the slope of $(\pi \Omega c)$ in relation to c $\bar{x} = [1 - (c/c_0)^{-3/2}]^2$. It is measured without the knowledge of any material property and thermal shock conditions. In order to estimate the thermal shock problem, it is necessary to measure the cracks lengths only before and after thermal shock. Figure 25 represents the results of crack length measurements for alumina and FGM samples after thermal shock. One can notice that the parameter R_m is much higher for FGM than for Al_2O_3, which means that the thermal shock resistance of FGM is much better than that of pure alumina.

Calculated values of σth, on the basis of experiments, (according to (125) and (126)) can be compared to the theoretical estimate of the maximum

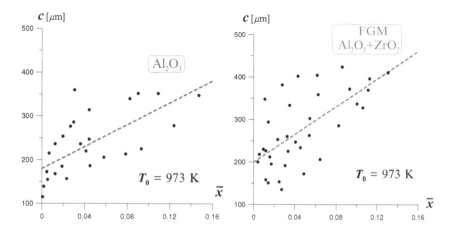

Figure 25.   Crack lengths after thermal shock in Al_2O_3 and in FGM

Table 1.   Characteristic material parameters defining the thermal shock process

Material	K_{Ic} (MPam$^{1/2}$)	R_m (mm)	σth (MPa)	$\bar{\sigma}_{rr}^{\max}(r = 0, z = H)$ (MPa)
Al$_2$O$_3$	3.36	1.64	83	84
FGM	3.36	2.06	74	77

$\bar{\sigma}_{rr}^{\max}(r = 0, z = H)$ (Fig. 24). The value of K_{Ic} in (125) was assessed in a 4-point bending test with Vicker's indentations. The characteristic calculation for the thermal shock starting at $T_0 = 973$ K are collected in Table 1. They indicate that both values of stress are very similar.

In conclusion one can say this method is effective for the description of crack propagation under 2-D non-symmetric thermal shock.

9.   Concluding remarks

We have proposed methodology for describing CMC and FGM thermomechanical response when subjected to sudden changes of temperature. We discussed the modelling of thermomechanical properties both for ceramic matrix and metal matrix composites. Appropriate gradation of composite properties can significantly improve thermal shock response of the material itself or a structural element. We classified different approaches for different types of composites. We formulated the governing equations applied for the temperature field under transient thermal loading and solved them numerically by using FEA and FD methods. The role of thermal residual stresses is significant in the analysis of the thermal shock problems. In a non-symmetric gradation profile they create the initial curvature of the FGM structural element. The basic fracture mechanics ideas were presented for CMC and FGM subjected to a transient temperature field. The crack-bridging mechanism plays an important role in the assessment of the response of the composites. Numerical examples dealing with 1-D and 2-D non-symmetric thermal shock problem illustrated the effectiveness of the theoretical methods to the solution of practical problems.

Acknowledgement

The author was supported by grant No 65/6.PR UE/2005-2008/7 of the Polish Ministry of Science and Higher Education. The author wishes to express his gratitude to Marie Curie Transfer of Knowledge programm for the support of project MKTD-CT-2004-014058.

References

J. Aboudi, *Mechanics of Composite Materials*, Elsevier, Amsterdam, 1991.

K.E. Aifantis, J.R. Willis, *Mech. Mater.*, 2006, **38**, 702–716.

K.E. Aifantis, J.R. Willis, *J. Mech. Phys. Solids*, 2005, 53, 1047–70.

A. Atkinson, A. Selçuk, *Acta Mater.*, 1999, **47**, 867–74.

T.L. Becker, R.M. Cannon, R.O. Ritchie, *Mech. Mater.*, 2000, **32**, 85–97.

Y. Benveniste, *Mech. Mater.*, 1987, **6**, 147–157.

T.J. Chung, A. Neubrand, J. Rödel, T. Fett, *Ceram. Trans.*, 2001, **114**, 789–96.

M. Dao, P. Gu, A. Maewal, J. Asaro, *Acta Mater.*, 1997, **45**, 3265–76.

M. Dong, S. Schmauder, *Acta Mater.*, 1996, **44**, 2465–78.

F. Erdogan, *Com. Eng.*, 1995, **5**, 753–70.

F. Erdogan, B.H. Wu *J. Therm. Stress*, 1996, **19**, 237–65.

J.D. Eshelby, *Proc. Roy. Soc.*, 1957, **A 349**, 376–96.

A. Evans, *J. Am. Cer. Soc.*, 1990, **73**, 187–206.

Z. Fan, A.P. Miodownik, P. Tsakiropoulos, *Mat. Sci. Technol.*, 1992, **8**, 922–29.

Z. Fan, A.P. Miodownik, P. Tsakiropoulos, *Mat. Sci. Technol.*, 1993, **9**, 1094–100.

F. Felten, S. Schneider, T. Sadowski, *Int. J. Ref. Mat. Hard Mat.*, 2008, **26**, 55–60.

N.A. Fleck, J.R. Willis, *J. Mech. Phys. Solids*, 2004, **52**, 1855–88.

L.B. Freund, *J. Cryst. Growth*, 1993, **132**, 341–44.

L.B. Freund, *J. Mech. Phys. Solids*, 1996, **44**, 723–36.

L.B. Freund, *J. Mech. Phys. Solids*, 2000, **48**, 1159–74.

T. Fujimoto, N. Noda, *J. Am. Cer. Soc.*, 2001, **84**, 1480–86.

A.E. Gianakopoulos, S. Suresh, M. Finot, M. Olsson, *Acta Matall. Mater.*, 1995, **43**, 1335–54.

D. Gross, T. Seelig, *Fracture Mechanics with an Introduction to Micromechanics*, Springer, 2006.

Z. Hashin, S. Shtrikman, *J. Mech. Phys. Solids*, 1963, **11**, 127–40.

R. Hill, *J. Mech. Phys. Solids*, 1965, **13**, 213–22.

C.H. Hsueh, *Thin Sol. Films*, 2002a, **418**, 182–88.

C.H. Hsueh, *J. Appl. Phys.*, 2002b, **91**, 9652–56.

C.H. Hsueh, A.G. Evans, *J. Am. Cer. Soc.*, 1985, **68**, 241–48.

C.H. Hsueh, S. Lee, T.J. Chuang, *J. Appl. Mech.*, 2003a, **70**, 151–54.

C.H. Hsueh, S. Lee, *Composites B*, 2003b, **34**, 747–52.

M. Jiang, I. Jasiuk, M. Ostoja-Starzewski, *Comput. Mat. Sci.*, 2000, **25**, 329–38.

Z.-H. Jin, R.C. Batra, *J. Mech. Phys. Solids*, 1996a, **44**, 1221–35.

Z.-H. Jin, R.C. Batra, *J. Therm. Stress*, 1996b, **19**, 317–39.

Z.-H. Jin, R.C. Batra, *J. Therm. Stress*, 1998, **21**, 151–76.

Z.-H. Jin, G.H. Paulino, R.H. Dodds Jr., *Eng. Fract. Mech.*, 2003, **70**, 1885–912.

D. Jeulin, M. Ostoja-Starzewski (Eds), (2001) *Mechanics of Random and Multiscale Microstructures*, CISM Courses and Lectures no. 430, Wien, Springer, New York.

A.C. Kaya, F. Erdogan, *Quart. Appl. Math.*, 1987, **45**, 105–22.

A. Kawasaki, R. Watanabe, *Ceram. Trans.*, 1993, **34**, 157–64.

A. Kawasaki, R. Watanabe, *Compos. Part B*, 1997, **28**, 29–35.

A. Kawasaki, R. Watanabe, *Eng. Fract. Mech.*, 2002, **69**, 1713–28.

K.-S. Kim, N. Noda,, *Arch. Appl. Mech.*, 2002, **72**, 127–37.

M. Kouzeli, D.C. Dunand, *Acta Mater.*, 2003, **51**, 6105–21.

W. Kreher, W. Pomope, *Internal Stress in Heterogeneous Materials*, Akademie Verlag, Berlin, 1989.

V.D. Krstic, *Phil. Mag. A*, 1983, **48**, 695–708.

G. Li, P. Ponte Castañeda, *Appl. Mech. Rev.*, 1994, **47**, S77–94.

Z. Li, S. Schmauder, M. Dong, *Comput. Mater. Sci.*, 1999, **15**, 11–21.

M.N. Miller, *J. Math. Phys.*, 1969, **10**, 1988–2004.

R.J. Moon, M. Tilbrook, M. Hoffman, A. Neubrand, *J. Am. Cer. Soc.*, 2005, **88**, 666–74.

T. Mori, K. Tanaka, *Acta Metall.*, 1973, **21**, 571–74.

K.W. Morton, D.F. Mayers, *Numerical Solutions of Partial Differential Equations, An Introduction*, Cambridge Univ. Press, Cambridge, 2005.

H. Moulinec, P. Sequet, *Eur. J. Mech. A/Solids*, 2003, **22**, 751–70.

R.J. Munro, *J. Am. Cer. Soc.*, 2001, **84**, 1190–92.

D. Munz, T. Fett, *Ceramics: Mechanical Properties, Failure Behaviour, Materials Selection*, Springer, Berlin, 1999.

K. Nakonieczny and T. Sadowski, *Comput. Mater. Sci.*, 2008, (doi: 10.1016/j.commatsci.2008.08.0.19)

S. Nemat-Nasser, M. Hori, *Micromechanics: Overall Properties of Heterogeneous Materials*, Nordh-Holland, Amsterdam, 1993.

A. Neubrand, J. Rödel, *Z. Metallkd.*, 1997, **88**, 358–71.

A. Neubrand, T.J. Chung, J. Rödel, E.D. Steffler, T. Fett, *J. Mat. Res.*, 2002, **17**, 2912–20.

N. Noda, *J. Thermal Stress*, 1999, **22**, 477–512.

N. Noda, R.B. Hetnarski, Y. Tanigawa, *Thermal stress*, Lastran Corporation, USA, 2000.

M. Ostoja-Starzewski, *Appl. Mech. Rev.*, 1994, **47**, S221–S230.

M. Ostoja-Starzewski, *Microstructural Randomness and Scaling in Mechanics of Materials*, Chapman&Hall/CRC, 2007.

M. Ostoja-Starzewski, J. Schulte, *Phys. Rev. B*, 1996, **54**, 278–85.

M. Ostoja-Starzewski, P.Y. Sheng, I. Jasiuk, *ASME J. of Engng. Mat. Tech.*, 1994, **116**, 384–91.

G.A. Papadopoulos, *Fracture Mechanics. The Experimental Method of Caustics and the Det.-Criterion of Fracture*, Springer, 1993.

M.J. Pindera, J. Aboudi, S.M. Arnold, *J. Am. Cer. Soc.*, 1998, **81**, 1525–36.

M.J. Pindera, J. Aboudi, S.M. Arnold, *Eng. Fract. Mech.*, 2002, **69**, 1587–606.

S. Pitakthapanaphong, E.P. Busso, *J. Mech. Phys. Solids*, 2002, **50**, 695–716.

P. Ponte Castañeda, *J. Mech. Phys. Solids*, 1991, **39**, 45–71.

P. Ponte Castañeda, P. Sequet, *Adv, in Appl. Mech.*, 1998, **34**, 171–302.

E. Postek,, T. Sadowski, S.J. Hardy, 2005. The mechanical response of a ceramic polycrystalline material with inter-granular layers. In: *Proc. VIII International Conference on Computational Plasticity, COMPLAS VIII*, E. Oñate, D.R.J. Owen eds., CIMNE, Barcelona, pp. 1075–78.

R. Pyrz, *Comp. Sci. Techn.*, 1994, **50**, 539–46.

R. Pyrz and B. Bochenek, *Sci. Engng. Comp. Mat.*, 1994, **3**, 95–109.

R. Pyrz and B. Bochenek, *Int. J. Solids Struct.*, 1998, **35**, 2413–27.

K.S. Ravichandran, *J. Am. Cer. Soc.*, 1994, **77**, 1178–84.

K.S. Ravichandran, *Mat. Sci. Eng.*, 1995, **A201**, 269–76.

T. Reiter, G.J. Dvorak, V. Tvergaard, *J. Mech. Phys. Solids*, 1997, **45**, 1281–302.

T. Sadowski, *Mech. Mater.*, 1994, **18**, 1–16.

T. Sadowski (Ed.), (2005a) *Multiscale Modelling of Damage and Fracture Processes in Composite Materials, CISM Courses and Lectures no. 474*, Wien, Springer, New York.

T. Sadowski, Modelling of Damage and Fracture Processes of Ceramic Matrix Composites. In *CISM Courses and Lectures No. 474*, ed. T. Sadowski, Springer, Wien, New York, 2005b, pp. 271–309.

T. Sadowski, S. Samborski, *Comput. Mater. Sci.*, 2003a, **28**, 512–17.

T. Sadowski, S. Samborski, *J. Am. Ceram. Soc.*, 2003b, **86**, 2218–21.

T. Sadowski, A. Neubrand, *Int. J. Frac.* 2004, **127**, L135–L140.

T. Sadowski T., A. Nowicki, *Comput. Mater. Sci.*, 2008, **43**, 235–41.

T. Sadowski, K. Nakonieczny, *Comput. Mater. Sci.*, 2008, **43**, 171–78.

T. Sadowski, S. J. Hardy, E. Postek, *Comput. Mater. Sci.*, 2005a, **34**, 46–63.

T. Sadowski, S. Samborski, Z. Librant, *Key Eng. Mat.*, 2005b, **290**, 86–93.

T. Sadowski, M. Boniecki, Z. Librant, *Adv. Sci. Tech.*, 2006a, **45**, 1640–45.

T. Sadowski, S. J. Hardy, E. Postek, *Mater. Sci. Eng. A*, 2006b, **424**, 230–38.

T. Sadowski, M. Boniecki, Z. Librant, *Proc. ECERS'07*, Berlin 17–21 June 2007b. (CD-ROM)

T. Sadowski, E. Postek, Ch. Denis, *Comput. Mat. Sci.*, 2007c, **39**, 230–36.

T. Sadowski, S. Ataya, A. Nakonieczny, *Comput. Mater. Sci.* In: 2008a. (doi:10.1016/j.commatsci.2008.07.011)

T. Sadowski, M. Boniecki, Z. Librant, A. Nakonieczny, *Int. J. Heat Mass Transfer,* 2007a, **50**, 4461–67.

T. Sadowski, T. Niezgoda, K. Kosiuczenko, M. Boniecki, Z. Librant, 2008b, (in preparation)

S. Schmauder, U. Weber, I. Hofinger, A. Neubrand, *Tech. Mech.*, 1999, **19**, 313–20.

S. Schmauder, U. Weber, E. Soppa, *Comput. Mater. Sci.*, 2003, **26**, 142–53.

P. Suquet, (1997) *Continuum Micromechanics*, CISM Courses and Lectures no. 377, Wien, Springer, New York.

R. Stevens, *Introduction to Zirconia*, Magnesium Elektron Ltd., 1986.

S. Suresh, A. Mortensen, *Fundamentals of Functionally Graded Materials*, Institute of Materials, London, 1998.

D.R.S. Talbot, J.R. Willis, *IMA J. Appl.Math.*, 1985, **35**, 39–54.

D.R.S. Talbot, J.R. Willis, *Int. J Solids Struct.*, 1992, **29**, 1981–87.

D.R.S. Talbot, J.R. Willis, *Proc. Roy. Soc.*, 1994, **A447**, 365–84.

D.R.S. Talbot, J.R. Willis, *Proc. Roy. Soc.*, 2004, **A460**, 2705–23

I. Tamura, Y. Tomota, H. Ozawa, In: *Proc. Third Int. Con. Strength Metal land Alloys*, Vol. 1., Cambridge: Institute of Metals, 1973, 611–15.

Y. Tanigawa, M. Matsumoto, T. Akai, *JSME Int. Journal, Series A,* 1997, **40**, 84–93.

S. Torquato, *Int. J. Sol. Struct.*, 1998, **35**, 2385–2406.

S. Torquato, *Random Heterogeneous Materials*, Springer, New York, 2002.

F. Trancert, F. Osterstock, *Scr. Mat.*, 1997, **37**, 443–447.

L.I. Tuschinskii, *Poroshk. Metall.,* 1983, **7**, 85 (translated in *Powder Metallurgy and Metall Ceramics*, 1983.

B.-L. Wang, Y.-W. Mai, *Int. J. Mech. Sci.*, 2005, **47**, 303–17.

B.-L. Wang, Y.-W. Mai, X.-H. Zhang, *Acta Mater.*, 2004, **52**, 4961–72.

L.D. Wegner, L.J. Gibson, *Int. J. Mech. Sci.*, 2000, **42**, 925–42.

E. Werner, T. Siegmund, H. Weinhandl, F.D. Fisher, *Appl. Mech. Rev.*, 1994, **47**, S231–S240.

D.S. Wilkinson, W. Pompe, M. Oeshner, *Prog. Mat. Sci.*, 2001, **46**, 379–405.

B-L. Wang, Y.-W. Mai, *Int. J.Mech. Sci.*, 2005, **47**, 303–317.

B-L. Wang, Y.-W. Mai, X-H. Zhang, *Acta Mater.*, 2004, **52**, 4961–72.

J.R. Willis, *J. Mech. Phys. Solids*, 1977, **25**, 185–202.

J.R. Willis, *Eur. J. Mech. A/Solids*, 2000, **19**, S165–S184.

O.C. Zienkiewicz, R.L. Taylor, *The Finie Element Method*, McGraw-Hill, New York, 1987.

J.R. Zuiker, *Com. Eng.*, 1995, **7**, 807–19.

A PRECIS OF TWO-SCALE APPROACHES FOR FRACTURE IN POROUS MEDIA

R. de Borst
Department of Mechanical Engineering
Eindhoven University of Technology
NL-5600 MB Eindhoven
The Netherlands &
LaMCoS, UMR CNRS 5514
INSA de Lyon
Villeurbanne, France
r.d.borst@tue.nl

J. Rethore
LMT, E.N.S. de Cachan, Cachan, France

M.-A. Abellan
LTDS, UMR CNRS 5513, ENISE, Saint-Etienne, France

Abstract   A derivation is given of two-scale models that are able to describe deformation and flow in a fluid-saturated and progressively fracturing porous medium. From the micromechanics of the flow in the cavity, identities are derived that couple the local momentum and the mass balances to the governing equations for a fluid-saturated porous medium, which are assumed to hold on the macroscopic scale. By exploiting the partition-of-unity property of the finite element shape functions, the position and direction of the fracture is independent from the underlying discretisation. The finite element equations are derived for this two-scale approach and integrated over time. The resulting discrete equations are nonlinear due to the cohesive crack model and the nonlinearity of the coupling terms. A consistent linearisation is given for use within a Newton–Raphson iterative procedure. Finally, examples are given to show the versatility and the efficiency of the approach.

Keywords: multiscale analysis, partition-of-unity approach, fracture, porous media, multiphase media, cohesive cracks

R. de Borst and T. Sadowski (eds.) *Lecture Notes on Composite Materials – Current Topics and Achievements*

1.   Introduction

Since the pioneering work of Terzaghi [1] and Biot [2] the flow of fluids in deforming porous media has received considerable attention. Recently, Lewis and Schrefler [3] have given an account of this topic which is crucial for understanding and predicting the physical behaviour of many systems of interest, for example, in geotechnical and petroleum engineering, but also for soft tissues. Because of the complicated structure and functioning of human tissues, the classical two-phase theory has been extended to three and four-phase media, taking into account ion transport and electrical charges [4–6]. A general framework for accommodating multi-field problems has been presented by Jouanna and Abellan [7].

In spite of the importance of the subject, flow in *damaged* porous media has received little attention. Yet, the presence of damage, such as cracks, faults, and shear bands, can markedly change the physical behaviour. Furthermore, the fluid can transport contaminants which can dramatically reduce the strength of the solid skeleton. To account for such phenomena, the fluid flow must be studied also in the presence of discontinuities in the solid phase. The physics of the flow within such discontinuities can be very different from that of the interstitial fluid in the deforming bulk material. These differences affect the flow pattern and therefore also the deformations in the vicinity of the discontinuity. As we will show at the end of the paper, the local differences in flow characteristics can even influence the flow and deformations in the entire body of interest.

In this contribution, we will describe a general numerical methodology to capture deformation and flow in progressively fracturing porous media, summarising and unifying the recent work reported in [8–11]. The model allows for flow inside an evolving crack to be in the tangential direction. This is achieved by a priori adopting a two-scale approach. At the fine scale the flow in the cavity created by the (possibly cohesive) crack can be modelled in various ways, e.g. as a Stokes flow in an open cavity, or using a Darcy relation for a damaged porous material. Since the cross-sectional dimension of the cavity is small compared to its length, the flow equations can be averaged over the width of the cavity. The resulting equations provide the momentum and mass couplings to the standard equations for a porous medium, which are assumed to hold on the macroscopic scale.

Numerically, the two-scale model which ensues, imposes some requirements on the interpolation of the displacement and pressure fields near the discontinuity. The displacement field must be discontinuous across the cavity. Furthermore, the micromechanics of the flow within the cavity require that the flow normal to the cavity is discontinuous, and in conformity with Darcy's relation which, at the macroscopic scale, is assumed to hold for the surrounding porous medium, the normal derivative of the fluid pressure field must also be discontinuous from one face of the cavity to the other. For arbitrary discretisations, these requirements can be satisfied by exploiting the partition-of-unity property of finite element shape functions [12], as has been done successfully in applications to cracking in single-phase media [13–23].

To provide a proper setting, we will first briefly recapitulate the governing equations for a deforming porous medium under quasi-static loading conditions. The strong as well as the weak formulations will be considered, since the latter formulation is crucial for incorporating the micromechanical flow model properly. This micromechanical flow model is discussed in the next section, where it will be demonstrated how the momentum and mass couplings of the micromechanical flow model to the surrounding porous medium can be accomplished in the weak formulation. Time integration and a consistent linearisation of the resulting equations, which are nonlinear due to the coupling terms and the cohesive crack model complete the numerical model. Finally, example calculations are given of a body with stationary and propagating cracks.

2. Balance equations

We consider a two-phase medium subject to the restriction of small displacement gradients and small variations in the concentrations [7]. Furthermore, the assumptions are made that there is no mass transfer between the constituents and that the processes which we consider, occur isothermally. With these assumptions, the balances of linear momentum for the solid and the fluid phases read

$$\nabla \cdot \boldsymbol{\sigma}_\pi + \hat{\mathbf{p}}_\pi + \rho_\pi \mathbf{g} = \frac{\partial(\rho_\pi \mathbf{v}_\pi)}{\partial t} + \nabla \cdot (\rho_\pi \mathbf{v}_\pi \otimes \mathbf{v}_\pi) \qquad (1)$$

with $\boldsymbol{\sigma}_\pi$ the stress tensor, ρ_π the apparent mass density, and $\mathbf{v}_\pi$ the absolute velocity of constituent π. As in the remainder of this paper, $\pi = s, f$, with s and f denoting the solid and fluid phases, respectively. Further, $\mathbf{g}$ is the gravity acceleration and $\hat{\mathbf{p}}_\pi$ is the source of momentum for constituent π from the other constituent, which takes into account the possible local drag interaction between the solid and the fluid. Evidently, the latter source terms must satisfy the momentum production constraint:

$$\sum_{\pi=s,f} \hat{\mathbf{p}}_\pi = \mathbf{0} \qquad (2)$$

We now neglect convective, gravity and acceleration terms, so that the momentum balances reduce to

$$\nabla \cdot \boldsymbol{\sigma}_\pi + \hat{\mathbf{p}}_\pi = \mathbf{0} \qquad (3)$$

Adding both momentum balances, and taking into account Eq. (2), one obtains the momentum balance for the mixture:

$$\nabla \cdot \boldsymbol{\sigma} = \mathbf{0} \tag{4}$$

where the stress is, as usual, composed of a solid and a fluid part, $\boldsymbol{\sigma} = \boldsymbol{\sigma}_s + \boldsymbol{\sigma}_f$.

In a similar fashion as for the balances of momentum, one can write the balance of mass for each phase as

$$\frac{\partial \rho_\pi}{\partial t} + \nabla \cdot (\rho_\pi \mathbf{v}_\pi) = 0 \tag{5}$$

Again neglecting convective terms, the mass balances can be simplified to give

$$\frac{\partial \rho_\pi}{\partial t} + \rho_\pi \nabla \cdot \mathbf{v}_\pi = 0 \tag{6}$$

We multiply the mass balance for each constituent π by its volumetric ratio n_π, add them and utilise the constraint

$$\sum_{\pi=s,f} n_\pi = 1 \tag{7}$$

to give

$$\nabla \cdot \mathbf{v}_s + n_f \nabla \cdot (\mathbf{v}_f - \mathbf{v}_s) + \frac{n_s}{\rho_s}\frac{\partial \rho_s}{\partial t} + \frac{n_f}{\rho_f}\frac{\partial \rho_f}{\partial t} = 0 \tag{8}$$

The change in the mass density of the solid material is related to its volume change by

$$\nabla \cdot \mathbf{v}_s = -\frac{K_s}{K_t}\frac{n_s}{\rho_s}\frac{\partial \rho_s}{\partial t} \tag{9}$$

with K_s the bulk modulus of the solid material and K_t the overall bulk modulus of the porous medium. Using the Biot coefficient, $\alpha = 1 - K_t/K_s$ [3], this equation can be rewritten as

$$(\alpha - 1)\nabla \cdot \mathbf{v}_s = \frac{n_s}{\rho_s}\frac{\partial \rho_s}{\partial t} \tag{10}$$

For the fluid phase, a phenomenological relation is assumed between the incremental changes of the apparent fluid mass density and the fluid pressure p [3]:

$$\frac{1}{Q}\mathrm{d}p = \frac{n_f}{\rho_f}\mathrm{d}\rho_f \tag{11}$$

with the overall compressibility, or Biot modulus

$$\frac{1}{Q} = \frac{\alpha - n_f}{K_s} + \frac{n_f}{K_f} \tag{12}$$

where K_f is the bulk modulus of the fluid. Inserting relations (10) and (11) into the balance of mass of the total medium, Eq. (8), gives

$$\alpha \nabla \cdot \mathbf{v}_s + n_f \nabla \cdot (\mathbf{v}_f - \mathbf{v}_s) + \frac{1}{Q} \frac{\partial p}{\partial t} = 0 \tag{13}$$

The field equations, i.e. the balance of momentum of the saturated medium, Eq. (4), and the balance of mass, Eq. (13), are complemented by the boundary conditions

$$\mathbf{n}_\Gamma \cdot \boldsymbol{\sigma} = \mathbf{t}_p, \qquad \mathbf{v}_s = \mathbf{v}_p \tag{14}$$

which hold on complementary parts of the boundary $\partial \Omega_t$ and $\partial \Omega_v$, with $\Gamma = \partial \Omega = \partial \Omega_t \cup \partial \Omega_v$, $\partial \Omega_t \cap \partial \Omega_v = \emptyset$, $\mathbf{t}_p$ being the prescribed external traction and $\mathbf{v}_p$ the prescribed velocity, and

$$n_f(\mathbf{v}_f - \mathbf{v}_s) \cdot \mathbf{n}_\Gamma = q_p, \qquad p = p_p \tag{15}$$

which hold on complementary parts of the boundary $\partial \Omega_q$ and $\partial \Omega_p$, with $\Gamma = \partial \Omega = \partial \Omega_q \cup \partial \Omega_p$ and $\partial \Omega_q \cap \partial \Omega_p = \emptyset$, q_p and p_p being the prescribed outflow of pore fluid and the prescribed pressure, respectively. The initial conditions specify the displacements $\mathbf{u}_\pi$, the velocities $\mathbf{v}_\pi$, and the pressure field at $t = 0$,

$$\mathbf{u}_\pi(\mathbf{x}, 0) = \mathbf{u}_\pi^0, \qquad \mathbf{v}_\pi(\mathbf{x}, 0) = \mathbf{v}_\pi^0, \qquad p(\mathbf{x}, 0) = p^0 \tag{16}$$

3. Constitutive equations

The effective stress increment in the solid skeleton, $\mathrm{d}\boldsymbol{\sigma}_s'$ is related to the strain increment $\mathrm{d}\boldsymbol{\epsilon}_s$ by an incrementally linear stress-strain relation for the solid skeleton,

$$\mathrm{d}\boldsymbol{\sigma}_s' = \bar{\mathbf{D}}^{tan} : \mathrm{d}\boldsymbol{\epsilon}_s \tag{17}$$

where $\bar{\mathbf{D}}^{tan}$ is the fourth–order tangent stiffness tensor of the solid material and the d symbol denotes a small increment. Since the effective stress in the solid skeleton is related to the partial stress by $\boldsymbol{\sigma}_s' = \boldsymbol{\sigma}_s/n_s$, the above relation can be replaced by

$$\mathrm{d}\boldsymbol{\sigma}_s = \mathbf{D}^{tan} : \mathrm{d}\boldsymbol{\epsilon}_s \tag{18}$$

where the notation $\mathbf{D}^{tan} = n_s \bar{\mathbf{D}}^{tan}$ has been used. In the examples, a linear-elastic behaviour of the bulk material has been assumed, and we have set $\mathbf{D}^{tan} = \mathbf{D}$, the linear-elastic stiffness tensor.

At the discontinuity Γ_d a discrete relation holds between the interface tractions $\mathbf{t}_d$ and the relative displacements $\boldsymbol{\delta}$:

$$\mathbf{t}_d = \mathbf{t}_d(\boldsymbol{\delta}, \kappa) \tag{19}$$

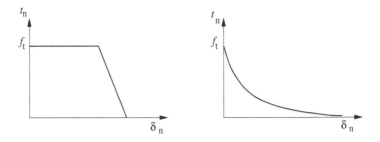

Figure 1. Decohesion curves for a ductile (left) and a quasi-brittle solid (right)

with κ a history parameter. After linearisation, necessary to use a tangential stiffness matrix in an incremental-iterative solution procedure, one obtains

$$\dot{\mathbf{t}}_d = \mathbf{T}\dot{\boldsymbol{\delta}} \tag{20}$$

with $\mathbf{T}$ the material tangent stiffness matrix of the discrete traction-separation law:

$$\mathbf{T} = \frac{\partial \mathbf{t}_d}{\partial \boldsymbol{\delta}} + \frac{\partial \mathbf{t}_d}{\partial \kappa}\frac{\partial \kappa}{\partial \boldsymbol{\delta}} \tag{21}$$

A key element is the presence of a work of separation or fracture energy, $\mathcal{G}_c$, which governs crack growth and enters the interface constitutive relation (19) in addition to the tensile strength f_t. It is defined as the work needed to create a unit area of fully developed crack:

$$\mathcal{G}_c = \int_{\delta_n=0}^{\infty} \sigma \mathrm{d}\delta_n \tag{22}$$

with σ the stress across the fracture process zone. It thus equals the area under the decohesion curves as shown in Fig. 1. Evidently, cohesive-zone models as defined above are equipped with an internal length scale, since the quotient $\mathcal{G}_c/E$, with E a stiffness modulus for the surrounding continuum, has the dimension of length.

In a standard manner we adopt Darcy's relation for the flow in the bulk of the porous medium,

$$n_f(\mathbf{v}_f - \mathbf{v}_s) = -k_f \nabla p \tag{23}$$

with k_f the permeability coefficient. The equations for the flow in the cavity, which close the initial value problem, will be detailed in a subsequent section.

4.   Weak form of the balance equations

To arrive at the weak form of the balance equations, we multiply the momentum balance (4) and the mass balance (13) by kinetically admissible test functions for the displacements of the skeleton, $\boldsymbol{\eta}$, and for the pressure, ζ. After substitution of Darcy's relation for the flow in the porous medium, Eq. (23), integrating over the domain Ω and using the divergence theorem then leads to the corresponding weak forms:

$$\int_\Omega (\nabla \cdot \boldsymbol{\eta}) \cdot \boldsymbol{\sigma} \, d\Omega + \int_{\Gamma_d} [\![\boldsymbol{\eta} \cdot \boldsymbol{\sigma}]\!] \cdot \mathbf{n}_{\Gamma_d} \, d\Omega = \int_\Gamma \boldsymbol{\eta} \cdot \mathbf{t}_p \, d\Omega \qquad (24)$$

and

$$-\int_\Omega \alpha\zeta \nabla \cdot \mathbf{v}_s \, d\Omega + \int_\Omega k_f \nabla\zeta \cdot \nabla p \, d\Omega - \int_\Omega \zeta Q^{-1}\dot{p} \, d\Omega$$
$$+ \int_{\Gamma_d} \mathbf{n}_{\Gamma_d} \cdot [\![\zeta \, n_f(\mathbf{v}_f - \mathbf{v}_s)]\!] \, d\Gamma = \int_\Gamma \zeta \mathbf{n}_\Gamma \cdot \mathbf{q}_p \, d\Gamma \qquad (25)$$

Because of the presence of a discontinuity inside the domain Ω, the power of the external tractions on Γ_d and the normal flux through the faces of the discontinuity are essential features of the weak formulation. Indeed, these terms enable the momentum and mass couplings between the discontinuity – the microscopic scale – and the surrounding porous medium – the macroscopic scale.

The momentum coupling stems from the tractions across the faces of the discontinuity and the pressure applied by the fluid in the discontinuity onto the faces of the discontinuity. We assume stress continuity from the cavity to the bulk, so that we have

$$\boldsymbol{\sigma} \cdot \mathbf{n}_{\Gamma_d} = \mathbf{t}_d - p\mathbf{n}_{\Gamma_d} \qquad (26)$$

with $\mathbf{t}_d$ the cohesive tractions, which are given by Eq. (19). Therefore, the weak form of the balance of momentum becomes

$$\int_\Omega (\nabla \cdot \boldsymbol{\eta}) \cdot \boldsymbol{\sigma} \, d\Omega + \int_{\Gamma_d} [\![\boldsymbol{\eta}]\!] \cdot (\mathbf{t}_d - p\mathbf{n}_{\Gamma_d}) \, d\Gamma = \int_\Gamma \boldsymbol{\eta} \cdot \mathbf{t}_p \, d\Gamma \qquad (27)$$

Since the tractions have a unique value across the discontinuity, the pressure p must have the same value at both faces of the discontinuity, and, consequently, this must also hold for the test function for the pressure, ζ. Accordingly, the mass transfer coupling term for the water can be rewritten as follows:

$$-\int_\Omega \alpha\zeta \nabla \cdot \mathbf{v}_s \, d\Omega + \int_\Omega k_f \nabla\zeta \cdot \nabla p \, d\Omega - \int_\Omega \zeta Q^{-1}\dot{p} \, d\Omega$$
$$+ \int_{\Gamma_d} \zeta \mathbf{n}_{\Gamma_d} \cdot n_f [\![\mathbf{v}_f - \mathbf{v}_s]\!] \, d\Gamma = \int_\Gamma \zeta \mathbf{n}_\Gamma \cdot \mathbf{q}_p \, d\Gamma \qquad (28)$$

where $\mathbf{q}_d = n_f [\![\mathbf{v}_f - \mathbf{v}_s]\!]$ represents the fluid flux through the faces of the discontinuity.

The above identity for the coupling of the mass transfer can be interpreted as follows. Part of the fluid that enters the cavity through one of its faces flows away tangentially, that is in the cavity. Therefore, the fluid flow normal to the cavity is discontinuous. Because the fluid flow between the cavity and the surrounding porous medium has to be continuous at each of the faces of the discontinuity, and because the fluid velocity is related to the pressure gradient via Darcy's law, the gradient of the pressure normal to the discontinuity must be discontinuous across the crack.

5. Micro–macro coupling

To quantify the influence of the "micro"-flow inside the discontinuity on the "macro"-scale, the balances of mass and momentum at the microscale must be considered. Different assumptions can now be made for the cavity. First, we will assume an open cavity which is totally filled with a Newtonian fluid. Then, the balance of mass for the "micro"-flow in the cavity reads

$$\dot{\rho}_f + \rho_f \nabla \cdot \mathbf{v} = 0$$

subject to the assumptions of small changes in the concentrations and that convective terms can be neglected, cf. Eq. (6). We assume that the first term can be neglected because the problem is monophasic in the cavity and the velocities are therefore much higher than in the porous medium. With this assumption, and focusing on a two-dimensional configuration, the mass balance inside the cavity simplifies to

$$\frac{\partial v}{\partial x} + \frac{\partial w}{\partial y} = 0 \tag{29}$$

with $v = \mathbf{v} \cdot \mathbf{t}_{\Gamma_d}$ and $w = \mathbf{v} \cdot \mathbf{n}_{\Gamma_d}$ the tangential and normal components of the fluid velocity in the discontinuity, respectively, see Fig. 2. Accordingly, the difference in the fluid velocity components that are normal to both crack faces is given by

$$[\![w_f]\!] = -\int_{-h}^{h} \frac{\partial v}{\partial x} dy \tag{30}$$

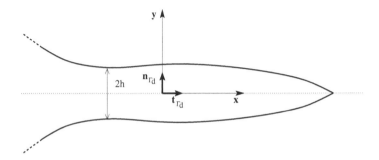

Figure 2.   Geometry and local coordinate system in cavity

To proceed, the velocity profile of the fluid flow inside the discontinuity must be known. From the balance of momentum for the fluid in the cavity and the assumption of a Newtonian fluid, the following velocity profile results

$$v(y) = \frac{1}{2\mu} \frac{\partial p}{\partial x}(y^2 - h^2) + v_f \tag{31}$$

where an integration has been carried out from $y = -h$ to $y = h$ and μ the viscosity of the fluid. The essential boundary $v = v_f$ has been applied at both faces of the cavity, and stems from the relative fluid velocity in the porous medium at $y = \pm h$:

$$v_f = (\mathbf{v}_s - n_f^{-1} k_f \nabla p) \cdot \mathbf{t}_{\Gamma_d}$$

Substitution of Eq. (31) into Eq. (30) and again integrating with respect to y then leads to

$$[\![w_f]\!] = \frac{2}{3\mu} \frac{\partial}{\partial x}\left(\frac{\partial p}{\partial x}h^3\right) - 2h\frac{\partial v_f}{\partial x} \tag{32}$$

This equation gives the amount of fluid attracted in the tangential fluid flow. It can be included in the weak form of the mass balance of the "macro"-flow to ensure the coupling between the "micro"-flow and the "macro"-flow. Since the difference in the normal velocity of both crack faces is given by $[\![w_s]\!] = 2\frac{\partial h}{\partial t}$, the mass coupling term becomes

$$\mathbf{n}_{\Gamma_d} \cdot \mathbf{q}_d = n_f[\![w_f - w_s]\!]$$
$$= n_f\left(\frac{2h^3}{3\mu}\frac{\partial^2 p}{\partial x^2} + \frac{2h^2}{\mu}\frac{\partial p}{\partial x}\frac{\partial h}{\partial x} - 2h\frac{\partial v_f}{\partial x} - 2\frac{\partial h}{\partial t}\right) \tag{33}$$

Another possibility for closure of the initial value problem is to assume that the cavity is partially filled with solid material. The mass balance for the fluid inside the cavity then reads

$$\alpha \nabla \cdot \mathbf{v}_s + n_f \nabla \cdot (\mathbf{v}_f - \mathbf{v}_s) + \frac{1}{Q}\frac{\partial p}{\partial t} = 0$$

Because the width of the cavity $2h$ is negligible compared to its length, the mass balance is again enforced in an average sense over the cross section. For the first term, we obtain

$$\int_{-h}^{h} \alpha \nabla \cdot \mathbf{v}_s dy = \int_{-h}^{h} \alpha\left(\frac{\partial v_s}{\partial x} + \frac{\partial w_s}{\partial y}\right)dy = \int_{-h}^{h} \alpha\frac{\partial v_s}{\partial x}dy - \alpha[\![w_s]\!] \tag{34}$$

where v_s and w_s are the component of the solid velocity tangential and normal to the crack, respectively. Because α depends on the capillarity pressure only, it can be assumed as constant over the cross section. Assuming furthermore that

v_s varies linearly with y, and defining $\langle v_s \rangle = \frac{1}{2}(v_s(h) + v_s(-h))$, the integral can be solved analytically

$$\int_{-h}^{h} \alpha \frac{\partial v_s}{\partial x} dy = 2\alpha h \left\langle \frac{\partial v_s}{\partial x} \right\rangle \tag{35}$$

Applying the same operations to the second term of Eq. (30), the following expression ensues

$$\int_{-h}^{h} n_f \nabla \cdot (\mathbf{v}_f - \mathbf{v}_s) dy = n_f [\![w_f - w_s]\!] + \frac{\partial}{\partial x} \left(\int_{-h}^{h} n_f (v_f - v_s) dy \right) - \\ \left(n_f(v_f(h) - v_s(h)) \frac{\partial h}{\partial x} - n_f(v_f(-h) - v_s(-h)) \frac{\partial(-h)}{\partial x} \right) \tag{36}$$

We now introduce Darcy's relation projected onto the axis tangential to the crack,

$$n_f(v_f - v_s) = -k_d \frac{\partial p}{\partial x} \tag{37}$$

with k_d the permeability of the damaged, porous material inside the cavity. The dependence of the permeability inside the cavity, k_d, on y will be negligible compared to that on x and accordingly, k_d can be assumed not to depend on y. However, the decohesion inside the cavity will affect the permeability, and therefore $k_d = k_d(h)$. Upon substitution of Eq. (37) into Eq. (36), the following relation is obtained

$$\int_{-h}^{h} n_f \nabla \cdot (\mathbf{v}_f - \mathbf{v}_s) dy = n_f [\![w_f - w_s]\!] - \frac{\partial}{\partial x} \left(\int_{-h}^{h} k_d \frac{\partial p}{\partial x} dy \right) + \\ \left(k_d \frac{\partial p(h)}{\partial x} \frac{\partial h}{\partial x} - k_d \frac{\partial p(-h)}{\partial x} \frac{\partial(-h)}{\partial x} \right) \tag{38}$$

At the faces of the cavity the permeability equals that of the bulk, and for continuity reasons $\frac{\partial p(h)}{\partial x} = \frac{\partial p(-h)}{\partial x}$, so that the second term becomes

$$\int_{-h}^{h} n_f \nabla \cdot (\mathbf{v}_f - \mathbf{v}_s) dy = n_f [\![w_f - w_s]\!] - 2h \frac{\partial k_d(h)}{\partial x} \frac{\partial p}{\partial x} - 2k_d h \frac{\partial^2 p}{\partial x^2} \tag{39}$$

For the third term, neglecting variations of the pressure over the height of the cavity, one obtains

$$\int_{-h}^{h} \frac{1}{Q} \frac{\partial p}{\partial t} dy = \frac{2h}{Q} \frac{\partial p}{\partial t} \tag{40}$$

Recalling that the term $n_f[\![w_f - w_s]\!]$ can be identified as the coupling term $\mathbf{n}_{\Gamma_d} \cdot \mathbf{q}_d$ of the weak form of the mass balance, we obtain

$$\mathbf{n}_{\Gamma_d} \cdot \mathbf{q}_d = -\frac{2h}{Q}\frac{\partial p}{\partial t} + 2\alpha\frac{\partial h}{\partial t} - 2\alpha h\langle\frac{\partial v_s}{\partial x}\rangle + 2h\frac{\partial k_d(h)}{\partial x}\frac{\partial p}{\partial x} + 2k_d h\frac{\partial^2 p}{\partial x^2} \quad (41)$$

6.   Discontinuities in a two-phase medium

A finite element method that can accommodate the propagation of disconti-nuities through elements was proposed by Belytschko and co-workers [13,14], exploiting the partition-of-unity property of finite element shape functions [12]. Since finite element shape functions φ_j form partitions of unity, $\sum_{j=1}^{n} \varphi_j = 1$ with n the number of nodal points, the components v_i of a velocity field $\mathbf{v}$ can be interpolated as

$$v_i = \sum_{j=1}^{n} \varphi_j \left(\dot{\bar{a}}_j + \sum_{k=1}^{m} \psi_k \dot{\hat{a}}_{jk} \right) \quad (42)$$

with $\bar{a}_j$ the "regular" nodal degrees-of-freedom for the displacements, ψ_k the enhanced basis terms, and $\hat{a}_{jk}$ the additional displacement degrees-of-freedom at node j which represent the amplitude of the kth enhanced basis term ψ_k. Next, we consider a domain Ω that is crossed by a single discontinuity at Γ_d (see Fig. 3). The velocity field $\mathbf{v}_s$ can then be written as the sum of two con-tinuous velocity fields $\bar{\mathbf{v}}_s$ and $\hat{\mathbf{v}}_s$:

$$\mathbf{v}_s = \bar{\mathbf{v}}_s + \mathcal{H}_{\Gamma_d}\hat{\mathbf{v}}_s \quad (43)$$

where $\mathcal{H}_{\Gamma_d}$ is the Heaviside step function centred at the discontinuity. The decomposition in Eq. (43) has a structure similar to the interpolation in Eq. (42). Accordingly, the partition-of-unity property of finite element shape functions enables the direct incorporation of discontinuities, including cracks and shear bands, in finite element models such that the discontinuous character of cracks

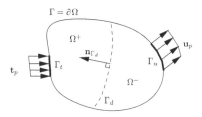

Figure 3.   Body composed of continuous displacement fields at each side of the discontinuity Γ_d

and shear bands is preserved. With the standard small-strain assumption that the strain-rate field of the solid, $\dot{\boldsymbol{\epsilon}}_s$, is derived from the symmetric part of the gradient of the velocity field, we obtain

$$\dot{\boldsymbol{\epsilon}}_s = \nabla^s \bar{\mathbf{v}}_s + \mathcal{H}_{\Gamma_d} \nabla^s \hat{\mathbf{v}}_s \tag{44}$$

away from the discontinuity Γ_d, with the superscript s denoting the symmetric part of the gradient operator.

With respect to the interpolation of the pressure p we note that the fluid flow normal to the discontinuity can be discontinuous. Since the fluid velocity is related to the pressure gradient via Darcy's relation, the gradient of the pressure normal to the discontinuity can therefore also be discontinuous across the discontinuity. Accordingly, the enrichment of the interpolation of the pressure must be such that the pressure itself is continuous, but has a discontinuous normal first spatial derivative. The distance function $\mathcal{D}_{\Gamma_d}$ defined as

$$\mathbf{n}_{\Gamma_d} \cdot \nabla \mathcal{D}_{\Gamma_d} = \mathcal{H}_{\Gamma_d} \tag{45}$$

satisfies this requirement, and accordingly, the interpolation of p will be such that

$$p = \bar{p} + \mathcal{D}_{\Gamma_d} \hat{p} \tag{46}$$

We now discretise the fields $\mathbf{v}_s$ and p and the test functions $\boldsymbol{\eta}$ and ζ for the velocity and the pressure, respectively, in a Bubnov-Galerkin sense:

$$\begin{aligned}
\mathbf{v}_s &= \mathbf{N}(\dot{\bar{\mathbf{a}}} + \mathcal{H}_{\Gamma_d} \dot{\hat{\mathbf{a}}}), & \boldsymbol{\eta} &= \mathbf{N}(\bar{\mathbf{w}} + \mathcal{H}_{\Gamma_d} \hat{\mathbf{w}}) \\
p &= \mathbf{H}(\bar{\mathbf{p}} + \mathcal{D}_{\Gamma_d} \hat{\mathbf{p}}), & \zeta &= \mathbf{H}(\bar{\mathbf{z}} + \mathcal{D}_{\Gamma_d} \hat{\mathbf{z}})
\end{aligned} \tag{47}$$

where the matrix $\mathbf{N}$ contains the shape functions N_i used as partition of unity for the interpolation of the velocity field $\mathbf{v}_s$, and $\mathbf{H}$ contains the shape functions H_i used as partition of unity for the interpolation of the pressure field p. $\dot{\bar{\mathbf{a}}}$ and $\dot{\hat{\mathbf{a}}}$ are the nodal arrays assembling the amplitudes that correspond to the standard and enhanced interpolations of the velocity field, and $\bar{\mathbf{p}}$ and $\hat{\mathbf{p}}$ assemble the amplitudes that correspond to the standard and enhanced interpolations of the pressure field. The choice for the interpolants in $\mathbf{N}$ and $\mathbf{H}$ is driven by modelling requirements. Indeed, the modelling of the fluid flow inside the cavity requires the computation of second derivatives of the pressure, see Eqs. (33) or (41). Hence, the order of the finite element shape functions H_i has to be sufficiently high, otherwise the coupling between the fluid flow in the cavity and the bulk will not be achieved. Further, the order of the finite element shape functions N_i must be greater than or equal to that of H_i for consistency in the discrete momentum balance.

Inserting Eq. (47) into Eqs. (27) and (28) and requiring that the result holds for all admissible $\bar{\mathbf{w}}$, $\bar{\mathbf{z}}$, $\hat{\mathbf{w}}$ and $\hat{\mathbf{z}}$ gives

$$\mathbf{f}_{\bar{a}}^{int} = \mathbf{f}_{\bar{a}}^{ext}$$
$$\mathbf{f}_{\hat{a}}^{int} = \mathbf{f}_{\hat{a}}^{ext}$$
$$\mathbf{f}_{\bar{p}}^{int} = \mathbf{f}_{\bar{p}}^{ext}$$
$$\mathbf{f}_{\hat{p}}^{int} = \mathbf{f}_{\hat{p}}^{ext}$$

(48)

with the external force vectors:

$$\mathbf{f}_{\bar{a}}^{ext} = \int_{\Gamma} \mathbf{N}^{\mathrm{T}} \mathbf{t}_p \mathrm{d}\Gamma, \qquad \mathbf{f}_{\hat{a}}^{ext} = \int_{\Gamma} \mathcal{H}_{\Gamma_d} \mathbf{N}^{\mathrm{T}} \mathbf{t}_p \mathrm{d}\Gamma$$
$$\mathbf{f}_{\bar{p}}^{ext} = \int_{\Gamma} \mathbf{H}^{\mathrm{T}} \mathbf{n}_{\Gamma}^{\mathrm{T}} \mathbf{q}_p \mathrm{d}\Gamma, \qquad \mathbf{f}_{\hat{p}}^{ext} = \int_{\Gamma} \mathcal{D}_{\Gamma_d} \mathbf{H}^{\mathrm{T}} \mathbf{n}_{\Gamma}^{\mathrm{T}} \mathbf{q}_p \mathrm{d}\Gamma$$

(49)

and the internal force vectors:

$$\mathbf{f}_{\bar{a}}^{int} = \int_{\Omega} \nabla \mathbf{N}^{\mathrm{T}} \boldsymbol{\sigma} \mathrm{d}\Omega$$
$$\mathbf{f}_{\hat{a}}^{int} = \int_{\Omega} \mathcal{H}_{\Gamma_d} \nabla \mathbf{N}^{\mathrm{T}} \boldsymbol{\sigma} \mathrm{d}\Omega + \int_{\Gamma_d} \mathbf{N}^{\mathrm{T}} (\mathbf{t}_d - p\mathbf{n}_{\Gamma_d}) \mathrm{d}\Gamma$$
$$\mathbf{f}_{\bar{p}}^{int} = -\int_{\Omega} \alpha \mathbf{H}^{\mathrm{T}} \mathbf{m}^{\mathrm{T}} \mathbf{v}_s \mathrm{d}\Omega + \int_{\Omega} k_f \nabla \mathbf{H}^{\mathrm{T}} \nabla p \mathrm{d}\Omega - \int_{\Omega} Q^{-1} \mathbf{H}^{\mathrm{T}} \dot{p} \mathrm{d}\Omega$$
$$\mathbf{f}_{\hat{p}}^{int} = -\int_{\Omega} \mathcal{D}_{\Gamma_d} \alpha \mathbf{H}^{\mathrm{T}} \mathbf{m}^{\mathrm{T}} \mathbf{v}_s \mathrm{d}\Omega + \int_{\Omega} k_f \nabla (\mathcal{D}_{\Gamma_d} \mathbf{H})^{\mathrm{T}} \nabla p \mathrm{d}\Omega$$
$$\quad - \int_{\Omega} \mathcal{D}_{\Gamma_d} Q^{-1} \mathbf{H}^{\mathrm{T}} \dot{p} \mathrm{d}\Omega + \int_{\Gamma_d} \mathbf{H}^{\mathrm{T}} \mathbf{n}_{\Gamma}^{\mathrm{T}} \mathbf{q}_d \mathrm{d}\Gamma$$

(50)

where, for two dimensions, $\mathbf{m}^{\mathrm{T}} = [1, 1, 0]$.

To carry out the time integration in (54) a backward finite difference scheme is adopted

$$\left(\frac{\mathrm{d}(.)}{\mathrm{d}t} \right)^{t+\Delta t} = \frac{(.)^{t+\Delta t} - (.)^{t}}{\Delta t} = \frac{\Delta(.)}{\Delta t}$$

(51)

where Δt is the time increment, while $(.)^{t}$ and $(.)^{t+\Delta t}$ denote the unknowns at t and $t + \Delta t$, respectively. Substitution of this identity into the set of semi-discrete Eq. (48) results in:

$$\mathbf{f}_{\bar{a}}^{int,t+\Delta t} = \mathbf{f}_{\bar{a}}^{ext,t+\Delta t}$$
$$\mathbf{f}_{\hat{a}}^{int,t+\Delta t} = \mathbf{f}_{\hat{a}}^{ext,t+\Delta t}$$
$$\mathbf{f}_{\bar{p}}^{int,t+\Delta t} = \Delta t \mathbf{f}_{\bar{p}}^{ext,t+\Delta t}$$
$$\mathbf{f}_{\hat{p}}^{int,t+\Delta t} = \Delta t \mathbf{f}_{\hat{p}}^{ext,t+\Delta t}$$

(52)

where

$$\mathbf{f}_{\bar{p}}^{int,t+\Delta t} = -\int_{\Omega} \alpha \mathbf{H}^{\mathrm{T}} \mathbf{m}^{\mathrm{T}} \mathbf{u}_s d\Omega + \int_{\Omega} k_f \nabla \mathbf{H}^{\mathrm{T}} \nabla p d\Omega \Delta t$$
$$- \int_{\Omega} Q^{-1} \mathbf{H}^{\mathrm{T}} \Delta p d\Omega$$

$$\mathbf{f}_{\hat{p}}^{int,t+\Delta t} = -\int_{\Omega} \mathcal{D}_{\Gamma_d} \alpha \mathbf{H}^{\mathrm{T}} \mathbf{m}^{\mathrm{T}} \mathbf{u}_s d\Omega + \int_{\Omega} k_f \nabla (\mathcal{D}_{\Gamma_d} \mathbf{H})^{\mathrm{T}} \nabla p d\Omega \, \Delta t$$
$$- \int_{\Omega} \mathcal{D}_{\Gamma_d} Q^{-1} \mathbf{H}^{\mathrm{T}} \Delta p d\Omega + \int_{\Gamma_d} \mathbf{H}^{\mathrm{T}} \mathbf{n}_{\Gamma}^{\mathrm{T}} \mathbf{q}_d d\Gamma \, \Delta t$$

while the other internal force vectors remain unchanged except that they are now evaluated at $t + \Delta t$.

For use in a Newton–Raphson solution method, the set (52) must be linearised. To this end, the stress and the pressure are decomposed as

$$\boldsymbol{\sigma}_j = \boldsymbol{\sigma}_{j-1} + \mathrm{d}\boldsymbol{\sigma}, \qquad p_j = p_{j-1} + \mathrm{d}p \tag{53}$$

with the subscripts $j-1$ and j signifying the iteration numbers. The linearisation of the set (52) then yields

$$\begin{bmatrix} \mathbf{K}_{\bar{a}\bar{a}} & \mathbf{K}_{\bar{a}\hat{a}} & \mathbf{C}_{\bar{a}\bar{p}} & \mathbf{C}_{\bar{a}\hat{p}} \\ \mathbf{K}_{\bar{a}\hat{a}} & \mathbf{K}_{\hat{a}\hat{a}} & \mathbf{C}_{\hat{a}\bar{p}} & \mathbf{C}_{\hat{a}\bar{p}} + \mathbf{C}_{\hat{a}\bar{p}}^* \Delta t \\ \mathbf{C}_{\bar{a}\bar{p}}^{\mathrm{T}} & \mathbf{C}_{\hat{a}\bar{p}}^{\mathrm{T}} & \mathbf{M}_{\bar{p}\bar{p}} + \Delta t \mathbf{K}_{\bar{p}\bar{p}} & \mathbf{M}_{\bar{p}\hat{p}} + \Delta t \mathbf{K}_{\bar{p}\hat{p}} \\ \mathbf{C}_{\bar{a}\hat{p}}^{\mathrm{T}} + \Delta t \mathbf{C}_{\bar{a}\hat{p}}^{+} & \mathbf{C}_{\hat{a}\hat{p}}^{\mathrm{T}} + \Delta t \mathbf{C}_{\hat{a}\hat{p}}^{+} & \mathbf{M}_{\bar{p}\hat{p}} + \Delta t \mathbf{K}_{\bar{p}\hat{p}}^{\mathrm{T}} + \Delta t \mathbf{K}_{\bar{p}\hat{p}}^{+} & \mathbf{M}_{\hat{p}\hat{p}} + \Delta t \mathbf{K}_{\hat{p}\hat{p}} \end{bmatrix} \begin{pmatrix} \mathrm{d}\bar{\mathbf{a}} \\ \mathrm{d}\hat{\mathbf{a}} \\ \mathrm{d}\bar{\mathbf{p}} \\ \mathrm{d}\hat{\mathbf{p}} \end{pmatrix} =$$

$$\times \begin{pmatrix} \mathbf{f}_{\bar{a}}^{ext,t+\Delta t} - \mathbf{f}_{\bar{a},j-1}^{int,t+\Delta t} \\ \mathbf{f}_{\hat{a}}^{ext,t+\Delta t} - \mathbf{f}_{\hat{a},j-1}^{int,t+\Delta t} \\ \Delta t \mathbf{f}_{\bar{p}}^{ext,t+\Delta t} - \mathbf{f}_{\bar{p},j-1}^{int,t+\Delta t} \\ \Delta t \mathbf{f}_{\hat{p}}^{ext,t+\Delta t} - \mathbf{f}_{\hat{p},j-1}^{int,t+\Delta t} \end{pmatrix} \tag{54}$$

with the stiffness matrices:

$$\mathbf{K}_{\bar{a}\bar{a}} = \int_{\Omega} \nabla \mathbf{N}^{\mathrm{T}} \mathbf{D} \nabla \mathbf{N} d\Omega$$

$$\mathbf{K}_{\bar{a}\hat{a}} = \int_{\Omega} \mathcal{H}_{\Gamma_d} \nabla \mathbf{N}^{\mathrm{T}} \mathbf{D} \nabla \mathbf{N} d\Omega$$

$$\mathbf{K}_{\hat{a}\hat{a}} = \int_{\Omega} \mathcal{H}_{\Gamma_d}^2 \nabla \mathbf{N}^{\mathrm{T}} \mathbf{D} \nabla \mathbf{N} d\Omega + \int_{\Gamma_d} \mathbf{N}^{\mathrm{T}} \mathbf{T} \mathbf{N} d\Gamma$$

$$\mathbf{K}_{\bar{p}\bar{p}} = -\int_{\Omega} k_f \nabla \mathbf{H}^{\mathrm{T}} \nabla \mathbf{H} d\Omega$$

$$\mathbf{K}_{\bar{p}\hat{p}} = -\int_{\Omega} k_f \nabla \mathbf{H}^{\mathrm{T}} \nabla (\mathcal{D}_{\Gamma_d} \mathbf{H}) \mathrm{d}\Omega$$

$$\mathbf{K}_{\hat{p}\hat{p}} = -\int_{\Omega} k_f \nabla (\mathcal{D}_{\Gamma_d} \mathbf{H})^{\mathrm{T}} \nabla (\mathcal{D}_{\Gamma_d} \mathbf{H}) \mathrm{d}\Omega \tag{55}$$

the mass matrices:

$$\mathbf{M}_{\bar{p}\bar{p}} = -\int_{\Omega} Q^{-1} \mathbf{H}^{\mathrm{T}} \mathbf{H} \mathrm{d}\Omega$$

$$\mathbf{M}_{\bar{p}\hat{p}} = -\int_{\Omega} \mathcal{D}_{\Gamma_d} Q^{-1} \mathbf{H}^{\mathrm{T}} \mathbf{H} \mathrm{d}\Omega$$

$$\mathbf{M}_{\hat{p}\hat{p}} = -\int_{\Omega} \mathcal{D}_{\Gamma_d}^2 Q^{-1} \mathbf{H}^{\mathrm{T}} \mathbf{H} \mathrm{d}\Omega \tag{56}$$

and the coupling matrices:

$$\mathbf{C}_{\bar{a}\bar{p}} = -\int_{\Omega} \alpha \nabla \mathbf{N}^{\mathrm{T}} \mathbf{m} \mathbf{H} \mathrm{d}\Omega$$

$$\mathbf{C}_{\bar{a}\hat{p}} = -\int_{\Omega} \alpha \mathcal{D}_{\Gamma_d} \nabla \mathbf{N}^{\mathrm{T}} \mathbf{m} \mathbf{H} \mathrm{d}\Omega$$

$$\mathbf{C}_{\hat{a}\bar{p}} = -\int_{\Omega} \alpha \mathcal{H}_{\Gamma_d} \nabla \mathbf{N}^{\mathrm{T}} \mathbf{m} \mathbf{H} \mathrm{d}\Omega$$

$$\mathbf{C}_{\hat{a}\bar{p}}^* = -\int_{\Gamma_d} \mathbf{N}^{\mathrm{T}} \mathbf{n}_{\Gamma_d}^{\mathrm{T}} \mathbf{H} \mathrm{d}\Gamma$$

$$\mathbf{C}_{\hat{a}\hat{p}} = -\int_{\Omega} \alpha \mathcal{H}_{\Gamma_d} \mathcal{D}_{\Gamma_d} \nabla \mathbf{N}^{\mathrm{T}} \mathbf{m} \mathbf{H} \mathrm{d}\Omega \tag{57}$$

The matrices $\mathbf{C}_{\bar{a}\hat{p}}^+$, $\mathbf{C}_{\hat{a}\hat{p}}^+$ and $\mathbf{K}_{\hat{p}\hat{p}}^+$ result from differentiating the mass coupling term $\int_{\Gamma_d} \mathbf{H}^{\mathrm{T}} \mathbf{n}_{\Gamma}^{\mathrm{T}} \mathbf{q}_d \mathrm{d}\Gamma$ with respect to $\bar{\mathbf{a}}$, $\hat{\mathbf{a}}$ and $\bar{\mathbf{p}}$, respectively. The resulting expressions depend on the subscale model that has been chosen for the microflow, and are generally very complicated. In any case, these coupling terms and the coupling term $\mathbf{C}_{\hat{a}\bar{p}}^*$ cause the tangent stiffness matrix to become unsymmetric. To restore symmetry, the non-symmetric contributions that arise from the mass and momentum couplings have been removed from the tangent stiffness matrix, and the iterations in the example calculations in the next section have been carried out with the resulting symmetric stiffness matrix.

7.   Examples: stationary and propagating cracks

First, an open, stationary crack is considered in a linear-elastic medium and no tractions are transferred between both sides of the crack. Hence, the first subscale model for the fluid flow in the cavity is used and singularity functions that stem from linear-elastic fracture mechanics are added as enrichment functions at the crack tips [10]. The specimen is a square-shaped fractured block

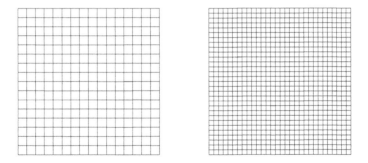

Figure 4.   Fine and coarse mesh composed of quadrilateral elements

under plane-strain conditions. A normal fluid flux $q_o = 10^{-4}$ ms^{-1} starting at $t = 0$ s is imposed at the bottom, while the top face is assigned a drained condition with zero pressure. Both left and right faces have undrained boundary conditions. No mechanical load is applied, but essential boundary conditions have been applied in order to remove rigid body motions. The block is 10 m × 10 m and consists of a porous material with a fluid volume fraction $n_f = 0.3$. The absolute mass densities are $\rho'_s = \rho_s/n_s = 2000$ kg/m^3 for the solid phase and $\rho'_f = \rho_f/n_f = 1000$ kg/m^3 for the fluid phase. The solid constituent is assumed to behave in a linear elastic manner with a Young's modulus $E = 9$ GPa and a Poisson's ratio $\nu = 0.4$. The Biot coefficient α has been set equal to 1, and the Biot modulus has been assigned a value $Q = 10^{18}$ GPa so as to simulate a quasi-incompressible fluid. This is not a limitation of the model, but more clearly brings out the influence of a fault. The bulk material has a permeability $k_f = 10^{-9}$ m^3/Ns.

Initially, one fault is considered. This fault, at the centre of the specimen, is 2 m long, and is inclined with the horizontal axis. A reference simulation has been run for a period of 10 s using 75 time steps. For this reference simulation, the coarse mesh shown in Fig. 4 has been used with quadrilateral elements equipped with quadratic shape functions.

The results obtained for an inclination angle of 30° are shown in Figs. 5–7. Because of the imposed fluid flux at the bottom, the pressure increases in the specimen and inside the fault, which subsequently opens, and fluid can flow inside, see Fig. 5. The discontinuity in the normal derivative of the pressure is illustrated in Fig. 6. Its jump has an opposite sign at both tips of the fault. The pore fluid flows into the cavity at the left tip where the gradient of the fault opening is positive. At the right tip, the gradient of the fault opening is negative and the fluid in the cavity flows back into the bulk. Figure 7 gives the absolute velocity field of the fluid. The "macro"-flow is oriented from the bottom to the top of the specimen. The "micro"-flow in the cavity is oriented by the direction

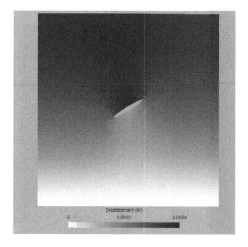

Figure 5. Deformed mesh for a crack angle of 30° (magnified ×1000)

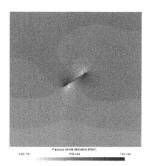

Figure 6. Normal derivative of the pressure $\mathbf{n}_{\Gamma_d} \cdot \nabla p$ for a crack angle of 30°. The top picture, where the data have been plotted as a piecewise-constant field, shows the results for the coarse mesh. Below, the results for the fine mesh have been smoothed and are depicted with filled iso-values

of the fault. In the cavity, the velocity of the fluid is very high compared with the velocity in the bulk because there is no resisting solid skeleton.

Next, a set of ten stationary faults is randomly generated. The length of the faults is between 1 and 3 m, while the fault angles vary from $-10°$ to $+30°$. Figure 8 shows the influence of the faults on the norm of the pressure gradient. The global fluid flow is strongly affected by the "micro"-flows inside the faults. From Fig. 8 it is observed that the main effect is due to the two longest faults. Evidently, the effect of a fault on the "macro"-flow increases with its length because more fluid can flow inside the cavity.

Finally, a cohesive crack is simulated which propagates in a plate under plane-strain conditions. Now, the second model for fluid flow inside the cavity has been used, in which progressive decohesion within the crack is assumed

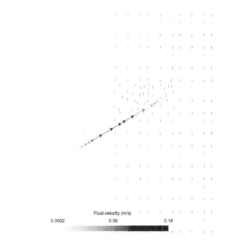

Figure 7.   Fluid velocity field for a crack angle of 30°

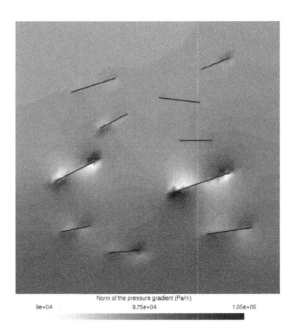

Figure 8. L_2 – Norm of the pressure gradient

[11]. The plate has sides of 0.25 m and an initial notch which is located at the symmetry axis and is 0.05 m deep. A fixed vertical velocity $v = 2.35 \times 10^{-2}$ μm/s is prescribed in a opposite direction at the bottom and at the top of the plate (tensile loading). All other boundaries of the plate are assumed to be impervious. The mesh consists of 20×20 quadrilateral elements with bilinear

Full coupling   No coupling

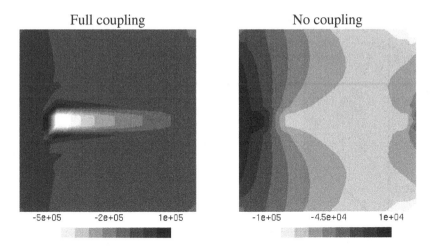

Figure 9. Pressure field in Pa. Left: The case with full coupling; Right: The case without coupling

shape functions for the pressure and the displacement fields. The time step size is 1 s and the analysis is continued until the crack reaches the right side of the plate.

Obviously, the amount of fluid that flows into the cavity is closely related to the crack opening displacement and vanishes at the crack tip. As a consequence the water pressure decreases in the vicinity of the crack. Figure 9 gives the pressure fields for a simulation with a full coupling at the interface as derived in the previous section and for a simulation without a mass transfer coupling term (and also without pressure enrichment). In the latter case, the crack is not seen as a discontinuity in the pressure field and the fluid phase flows through the crack as it does in the bulk. Accordingly, the pressure and pressure gradient are continuous at the interface. Indeed, the results presented in the two graphs of Fig. 9 are very different. Because of the mass transfer coupling, the water is sucked into the crack and high negative pressures occur around the crack. As a consequence, the water saturation decreases in this zone and intense cavitation occurs. Because of the negative values of the water pressure, sucking tractions modify the global response of the plate. As shown in Fig. 10, the load-displacement curve obtained with the coupling term results in a higher load-carrying capacity. Indeed, the effect of the mass transfer coupling is strong and changes the fluid flow in the entire domain. Moreover, the fluid velocity is increased by an order of magnitude.

Finally, the profile of the tangential component of the Darcian velocity is shown in Fig. 11. The graphs are plotted such that the velocity is positive when it is oriented from the actual crack tip to the initial pre-notch. When the coupling is activated, the water flows from the actual tip to the initial notch with a

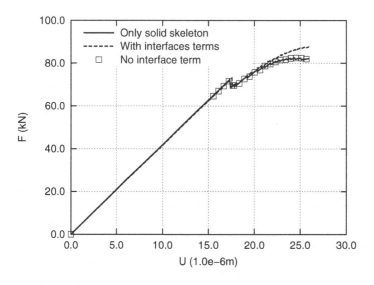

Figure 10.   Load-displacement curves

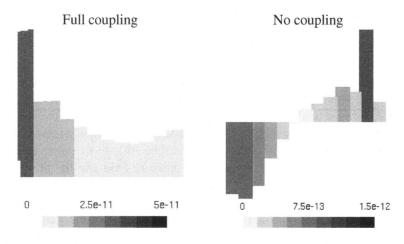

Figure 11.   Water tangential Darcian velocity profile inside the crack in m/s

value depending on the crack opening displacement whereas the orientation of the flow in the uncoupled case depends on the position on the interface.

8.   Concluding remarks

A methodology has been proposed to insert discontinuities such as cracks, faults, or shear bands, in a porous medium. The discontinuities can be located arbitrarily, not related to the underlying discretisation. For the fluid flow in the fractured porous medium a two-scale approach has been chosen, where the flow of the fluid inside the discontinuity (the "micro"-scale) is modelled independently from the flow of the pore fluid in the surrounding porous medium (the "macro"-scale). The mechanical and the mass transfer couplings between the two scales are obtained by inserting the homogenised "constitutive" relations of the "micro"-flow into the weak form of the balance equations of the bulk. The assumptions made for the fluid flow in and near the discontinuity require the addition of special enrichment functions for the displacement and the pressure fields. These conditions are satisfied by exploiting the partition of unity property of the finite element polynomial shape functions.

References

[1] Terzaghi K. *Theoretical Soil Mechanics*. Wiley, New York, 1943.

[2] Biot MA. *Mechanics of Incremental Deformations*. Wiley, Chichester, 1965.

[3] Lewis RW, Schrefler BA. *The Finite Element Method in the Static and Dynamic Deformation and Consolidation of Porous Media*, Second Edition. Wiley, Chichester, 1998.

[4] Snijders H, Huyghe JM, Janssen JD. Triphasic finite element model for swelling porous media. *International Journal for Numerical Methods in Fluids* 1997; **20:** 1039–1046.

[5] Huyghe JM, Janssen JD. Quadriphasic mechanics of swelling incompressible media. *International Journal of Engineering Science* 1997; **35:** 793–802.

[6] Van Loon R, Huyghe JM, Wijlaars MW, Baaijens FPT. 3D FE implementation of an incompressible quadriphasic mixture model. *International Journal for Numerical Methods in Engineering* 2003; **57:** 1243–1258.

[7] Jouanna P, Abellan MA. Generalized approach to heterogeneous media. In *Modern Issues in Non-Saturated Soils*. Gens A, Jouanna P, Schrefler B (eds). Springer, Wien, New York, 1995; 1–128.

[8] de Borst R, Réthoré J, Abellan MA. A numerical approach for arbitrary cracks in a fluid-saturated medium. *Archive of Applied Mechanics* 2006; **75:** 595–606.

[9] Réthoré J, de Borst R, Abellan MA. A discrete model for the dynamic propagation of shear bands in a fluid-saturated medium. *International Journal for Numerical and Analytical Methods in Geomechanics* 2007; **31:** 347–370.

[10] Réthoré J, de Borst R, Abellan MA. A two-scale approach for fluid flow in fractured porous media. *International Journal for Numerical Methods in Engineering* 2007; **71:** 780–800.

[11] Réthoré J, de Borst R, Abellan MA. A two-scale model for fluid flow in an unsaturated porous medium with cohesive cracks. *Computational Mechanics* 2008; **42:** 227–238.

[12] Babuska I, Melenk JM. The partition of unity method. *International Journal for Numerical Methods in Engineering* 1997; **40:** 727–758.

[13] Belytschko T, Black T. Elastic crack growth in finite elements with minimal remeshing. *International Journal for Numerical Methods in Engineering* 1999; **45:** 601–620.

[14] Moës N, Dolbow J, Belytschko T. A finite element method for crack growth without remeshing. *International Journal for Numerical Methods in Engineering* 1999; **46:** 131–150.

[15] Wells GN, Sluys LJ. Discontinuous analysis of softening solids under impact loading. *International Journal for Numerical and Analytical Methods in Geomechanics* 2001; **25:** 691–709.

[16] Wells GN, Sluys LJ, de Borst R. Simulating the propagation of displacement discontinuities in a regularized strain-softening medium. *International Journal for Numerical Methods in Engineering* 2002; **53:** 1235–1256.

[17] Wells GN, de Borst R, Sluys LJ. A consistent geometrically non-linear approach for delamination. *International Journal for Numerical Methods in Engineering* 2002; **54:** 1333–1355.

[18] Remmers JJC, de Borst R, Needleman A. A cohesive segments method for the simulation of crack growth. *Computational Mechanics* 2003; **31:** 69–77.

[19] Samaniego E, Belytschko T. Continuum–discontinuum modelling of shear bands. *International Journal for Numerical Methods in Engineering* 2005; **62:** 1857–1872.

[20] Réthoré J, Gravouil A, Combescure A. An energy-conserving scheme for dynamic crack growth using the extended finite element method. *International Journal for Numerical Methods in Engineering* 2005; **63:** 631–659.

[21] Réthoré J, Gravouil A, Combescure A. A combined space–time extended finite element method. *International Journal for Numerical Methods in Engineering* 2005; **64:** 260–284.

[22] Areias PMA, Belytschko T. Two-scale shear band evolution by local partition of unity. *International Journal for Numerical Methods in Engineering* 2006; **66:** 878–910.

[23] de Borst R, Remmers JJC, Needleman A. Mesh-independent numerical representations of cohesive-zone models. *Engineering Fracture Mechanics* 2006; **173:** 160–177.

INITIAL DEFORMATIONS ON BEHAVIOUR OF ELASTIC COMPOSITES

Eduard Marius Craciun

Faculty of Mathematics and Informatics
Ovidius University Constanta
blvd. Mamaia 124, cp 900527
Romania
mcraciun@univ-ovidius.ro

Abstract   Mathematical modelling of the incremental fields in a pre-stressed elastic composite in plane and antiplane states is done using complex variable theory. We formulate and solve the crack problem in all three classical modes by using complex potentials. Following Guz, and using the theory of Riemann-Hilbert problem, Cauchy's integral, Plemelj's functions we obtain the asymptotic behavior of the incremental fields in the vicinity of the crack tip. We obtain the critical values of the incremental stresses which produce crack propagation.

Keywords: pre-stressed elastic composite, crack, Riemann-Hilbert problem, complex potentials, resonance

1.   Homogeneous initial deformations

The fundamentals of the influence of homogeneous initial deformations are presented, for instance, in Guz's monographs [3, 4, 6] for elastic, viscoelastic, elastic-plastic and elastoviscoplastic materials; Ogden's monograph [9], analysed the case of elastic materials in great detail; Eringen's and Maugin's monographs [2] concerns the case of elastic materials interacting with electromagnetic and thermal fields. We present the basic results for hyperelastic materials; we follow Guz's approach exposed in the fundamental monograph [6] and in Cristescu et al. [8].

We assume that the initial applied deformations are *homogeneous*. Since the initial applied deformation is small, the fields can be considered as functions of $\mathbf{X}$ or of $\mathbf{x}$. We shall use the second variant. We denote the initial applied displacement field by $\overset{\circ}{\mathbf{u}} = \overset{\circ}{\mathbf{u}}(\mathbf{x})$, the incremental displacement field by

R. de Borst and T. Sadowski (eds.) *Lecture Notes on Composite Materials – Current Topics and Achievements*

$\mathbf{u} = \mathbf{u}(\mathbf{x})$, Cauchy's stress tensor, corresponding to the initial applied deformation by $\overset{\circ}{\boldsymbol{\sigma}} = \overset{\circ}{\boldsymbol{\sigma}}(\mathbf{x})$, the corresponding infinitesimal strain tensor by $\overset{\circ}{\boldsymbol{\varepsilon}} = \overset{\circ}{\boldsymbol{\varepsilon}}(\mathbf{x})$ and the incremental nominal stress tensor by $\boldsymbol{\theta} = \boldsymbol{\theta}(\mathbf{x})$.

The material is *linearly hyperelastic* (*physical linearity*) and the initial applied deformation is small (*geometrical linearity*).

The *incremental static behavior* of the body is governed by the following *incremental field equations*:

- The *incremental equilibrium equation*

$$\boldsymbol{\theta} = \mathbf{0} \quad \text{or} \quad \theta_{kl,k} = 0, \quad k, l = 1, 2, 3; \tag{1}$$

- The *incremental constitutive equation*

$$\boldsymbol{\theta} = \boldsymbol{\omega} \boldsymbol{\nabla} \mathbf{u}^{\mathrm{T}} \quad \text{or} \quad \theta_{kl} = \omega_{klmn} u_{m,n}, \quad k, l, m, n = 1, 2, 3; \tag{2}$$

  where the components of the instantaneous elasticity are given by

$$\omega_{klmn} = c_{klmn} + \overset{\circ}{\sigma}_{kn} \delta_{ml}, \tag{3}$$

- The *mixed incremental boundary conditions*

$$\mathbf{u} = \mathbf{v} \quad \text{or} \quad u_k = v_k \quad \text{on} \quad S_1 \quad \text{and}$$
$$\mathbf{s_n} = \boldsymbol{\theta}^{\mathrm{T}} \mathbf{n} = \mathbf{l} \quad \text{or} \quad s_{nl} = \theta_{kl} n_k = l_l \quad \text{on} \quad S_2. \tag{4}$$

The instantaneous elasticity $\boldsymbol{\omega}$ is *symmetric*; i.e.

$$\omega_{klmn} = \omega_{nmlk}. \tag{5}$$

The infinitesimal strain tensor $\overset{\circ}{\boldsymbol{\varepsilon}} = \overset{\circ}{\boldsymbol{\varepsilon}}(\mathbf{x})$, is given by

$$\overset{\circ}{\boldsymbol{\varepsilon}} = \frac{1}{2}\left(\boldsymbol{\nabla}\overset{\circ}{\mathbf{u}} + \boldsymbol{\nabla}\overset{\circ}{\mathbf{u}}^{\mathrm{T}}\right) \quad \text{or} \quad \overset{\circ}{\varepsilon}_{ml} = \frac{1}{2}\left(\overset{\circ}{u}_{m,l} + \overset{\circ}{u}_{l,m}\right). \tag{6}$$

The classical stress–strain relation is

$$\overset{\circ}{\boldsymbol{\sigma}} = \mathbf{c}\overset{\circ}{\boldsymbol{\varepsilon}} \quad \text{or} \quad \overset{\circ}{\sigma}_{kn} = c_{knml}\overset{\circ}{\varepsilon}_{ml}. \tag{7}$$

Initial applied deformation is small; i.e.

$$\left|\boldsymbol{\nabla}\overset{\circ}{\mathbf{u}}\right| \ll 1 \quad \text{or} \quad \left|\overset{\circ}{u}_{m,n}\right| \ll 1 \quad \text{for} \quad m, n = 1, 2, 3. \tag{8}$$

Cauchy's stress vector $\overset{\circ}{\mathbf{t}}_{\mathbf{n}}$ acting in the initial deformed configuration $\overset{\circ}{B}$ on a material surface element with unit normal $\mathbf{n}$ is given by the classical relation

$$\overset{\circ}{\mathbf{t}}_{\mathbf{n}} = \overset{\circ}{\boldsymbol{\sigma}}\mathbf{n} \quad \text{or} \quad \overset{\circ}{t}_{nk} = \overset{\circ}{\sigma}_{kl}n_l. \tag{9}$$

Actually, the latter equations can be used to determine the traction $\overset{\circ}{\mathbf{t}}_\mathbf{n}$ which must be applied on the boundary of the body in order to produce the initial displacement $\overset{\circ}{\mathbf{u}} = \overset{\circ}{\mathbf{u}}(\mathbf{x})$; i.e. the initial applied deformation of the material.

The Piola and Kirchhoff incremental stress vector $\mathbf{s_n}$ is given by

$$\mathbf{s_n} = \boldsymbol{\theta}^\mathrm{T}\mathbf{n} \quad \text{or} \quad s_{nl} = \theta_{kl}n_k. \tag{10}$$

We shall discuss the displacement potential problem.

Since ω_{klmn} are *constant* quantities, the incremental equation of equilibrium can be written in the following form:

$$P_{lm}u_m = 0, \tag{11}$$

where P_{lm} are differential operators defined by the equation

$$P_{lm} = \omega_{klmn}\frac{\partial^2}{\partial x_k \partial x_n}. \tag{12}$$

Introduce the 3×3 matrix

$$P = [P_{lm}] \tag{13}$$

and denote its determinant by $\det P$ and its inverse by P^{-1}.

We have the following relation from the matrix calculus

$$\frac{\partial \det P}{\partial P_{jm}} = \left(P^{-1}\right)_{mj}\det P. \tag{14}$$

We consider three regulated scalar valued functions

$$\varphi^{(j)} = \varphi^{(j)}(x_1, x_2, x_3). \quad j = 1, 2, 3. \tag{15}$$

Assume that the components $u_m^{(j)}$ of the incremental displacement $\mathbf{u}^{(j)}$ corresponding to $\varphi^{(j)}$ are expressed by

$$u_m^{(j)} = \frac{\partial \det P}{\partial P_{jm}}\varphi^{(j)}. \tag{16}$$

According with (12) we get

$$u_m^{(j)} = \left\{\left(P^{-1}\right)_{mj}\det P\right\}\varphi^{(j)}, \tag{17}$$

$$P_{lm}u_m^{(j)} = \left\{P_{lm}\left(P^{-1}\right)_{mj}\det P\right\}\varphi^{(j)} = (\delta_{lj}\det P)\,\varphi^{(j)}$$
$$= (\det P)\,\varphi^{(l)}. \tag{18}$$

Hence, the incremental equilibrium Eq. (1) will be satisfied by the incremental displacement fields $\mathbf{u}^{(j)}$ given in (16), if and only if the three functions $\varphi^{(j)}$ satisfy the same differential equation

$$(\det P)\,\varphi^{(j)} = 0 \quad \text{for} \quad j = 1, 2, 3. \tag{19}$$

The functions $\varphi^{(j)}$, giving $u^{(j)}$ through the relation (18) and satisfying the same differential Eq. (19), will be called *incremental displacement potentials* or briefly *displacement potentials*. The usefulness of the displacement potentials arises from the fact that all of them satisfy the same differential equation. In general, in the three-dimensional case (19), they satisfy a *sixth* order partial differential equation with constant coefficients. Moreover, even if the displacement potentials satisfy the same differential equation, in the boundary condition (2) all three potentials appear simultaneously and are *coupled*. We can say that even if we use the displacement potentials to solve incremental boundary value problems, the mathematical problem generally remains extremely complex.

Considerable simplification can be achieved only assuming supplementary symmetry properties of the material, particular forms for the initial imposed deformation; supplementary symmetry properties of the geometry of the body and particular forms of the imposed surface displacements and tractions.

For the symmetry properties of the material, we shall assume that the material is *orthotropic*, the symmetry planes being the coordinate planes. This assumption is wide enough for us to study the *stability behavior* of a large class of *composite materials* submitted to various loading conditions.

We shall assume that the initial imposed displacement $\overset{\circ}{\mathbf{u}}$ has the following components:

$$\overset{\circ}{u}_1 = (\lambda_1 - 1)x_1, \quad \overset{\circ}{u}_2 = (\lambda_2 - 1)x_2, \quad \overset{\circ}{u}_3 = (\lambda_3 - 1)x_3, \tag{20}$$

where $\lambda_1, \lambda_2, \lambda_3$ are given constant quantities. Since the initial deformation is small, $\lambda_1, \lambda_2, \lambda_3$ must satisfy the condition

$$|\lambda_k - 1| \ll 1 \quad \text{for} \quad k = 1, 2, 3. \tag{21}$$

According to (5) and (21) only the *diagonal* components of the initial infinitesimal strain $\overset{\circ}{\varepsilon}$ are nonvanishing, and we have

$$\overset{\circ}{\varepsilon}_{11} = \lambda_1 - 1, \quad \overset{\circ}{\varepsilon}_{22} = \lambda_2 - 1, \quad \overset{\circ}{\varepsilon}_{33} = \lambda_3 - 1, \quad \overset{\circ}{\varepsilon}_{kl} = 0 \quad \text{for} \quad k \neq l. \tag{22}$$

Since (21) holds, we have

$$\left|\overset{\circ}{\varepsilon}_{11}\right|, \left|\overset{\circ}{\varepsilon}_{22}\right|, \left|\overset{\circ}{\varepsilon}_{33}\right| \ll 1. \tag{23}$$

We conclude that only the *diagonal* components of the initial stress $\overset{\circ}{\sigma}$ are nonvanishing

$$\overset{\circ}{\sigma}_{11} = C_{11}\overset{\circ}{\varepsilon}_{11} + C_{12}\overset{\circ}{\varepsilon}_{22} + C_{13}\overset{\circ}{\varepsilon}_{33}, \quad \overset{\circ}{\sigma}_{22} = C_{21}\overset{\circ}{\varepsilon}_{11} + C_{22}\overset{\circ}{\varepsilon}_{22} + C_{23}\overset{\circ}{\varepsilon}_{33},$$

$$\overset{\circ}{\sigma}_{33} = C_{31}\overset{\circ}{\varepsilon}_{11} + C_{32}\overset{\circ}{\varepsilon}_{22} + C_{33}\overset{\circ}{\varepsilon}_{33}, \quad \overset{\circ}{\sigma}_{kl} = 0 \quad \text{for} \quad k \neq l.$$

$$(24)$$

Analysing the structure of the elasticity tensor **c** and using relations (3) and (24) we conclude that the nonvanishing components of the instantaneous elasticity $\boldsymbol{\omega}$ are given by

$$
\begin{aligned}
&\omega_{1111} = C_{11} + \overset{\circ}{\sigma}_{11}, \quad \omega_{2222} = C_{22} + \overset{\circ}{\sigma}_{22}, \quad \omega_{3333} = C_{33} + \overset{\circ}{\sigma}_{33}, \\
&\omega_{1122} = \omega_{2211} = C_{12} = C_{21}, \quad \omega_{2233} = \omega_{3322} = C_{23} = C_{32}, \\
&\omega_{3311} = \omega_{1133} = C_{13} = C_{31}, \\
&\omega_{1212} = \omega_{2121} = C_{66}, \quad \omega_{2323} = \omega_{3232} = C_{44}, \\
&\omega_{3131} = \omega_{1313} = C_{55}, \\
&\omega_{2112} = C_{66} + \overset{\circ}{\sigma}_{22}, \quad \omega_{1221} = C_{66} + \overset{\circ}{\sigma}_{11}, \\
&\omega_{3223} = C_{44} + \overset{\circ}{\sigma}_{33}, \quad \omega_{2332} = C_{44} + \overset{\circ}{\sigma}_{22}, \\
&\omega_{1331} = C_{55} + \overset{\circ}{\sigma}_{11}, \quad \omega_{3113} = C_{55} + \overset{\circ}{\sigma}_{33}.
\end{aligned}
$$

$$(25)$$

Simple computations lead to the following relations expressing the components of $\boldsymbol{\theta}$ in the terms of the components of $\nabla\mathbf{u}$:

$$
\begin{aligned}
\theta_{11} &= \omega_{1111}u_{1,1} + \omega_{1122}u_{2,2} + \omega_{1133}u_{3,3} \\
&= \left(C_{11} + \overset{\circ}{\sigma}_{11}\right)u_{1,1} + C_{12}u_{2,2} + C_{13}u_{3,3}, \\
\theta_{12} &= \omega_{1212}u_{1,2} + \omega_{1221}u_{2,1} = C_{66}u_{1,2} + \left(C_{66} + \overset{\circ}{\sigma}_{11}\right)u_{2,1}, \\
\theta_{13} &= \omega_{1313}u_{1,3} + \omega_{1331}u_{3,1} = C_{55}u_{1,3} + \left(C_{55} + \overset{\circ}{\sigma}_{11}\right)u_{3,1}, \\
\theta_{21} &= \omega_{2112}u_{1,2} + \omega_{2121}u_{2,1} = \left(C_{66} + \overset{\circ}{\sigma}_{22}\right)u_{1,2} + C_{66}u_{2,1}, \\
\theta_{22} &= \omega_{2211}u_{1,1} + \omega_{2222}u_{2,2} + \omega_{2233}u_{3,3} \\
&= C_{21}u_{1,1} + \left(C_{22} + \overset{\circ}{\sigma}_{22}\right)u_{2,2} + C_{23}u_{3,3}, \\
\theta_{23} &= \omega_{2323}u_{2,3} + \omega_{2332}u_{3,2} = C_{44}u_{2,3} + \left(C_{44} + \overset{\circ}{\sigma}_{22}\right)u_{3,2}, \\
\theta_{31} &= \omega_{3113}u_{1,3} + \omega_{3131}u_{3,1} = \left(C_{55} + \overset{\circ}{\sigma}_{33}\right)u_{1,3} + C_{55}u_{3,1}, \\
\theta_{32} &= \omega_{3223}u_{2,3} + \omega_{3232}u_{3,2} = \left(C_{44} + \overset{\circ}{\sigma}_{33}\right)u_{2,3} + C_{44}u_{3,2},
\end{aligned}
$$

$$(26)$$

$$\theta_{33} = \omega_{3311}u_{1,1} + \omega_{3322}u_{2,2} + \omega_{3333}u_{3,3}$$
$$= C_{31}u_{1,1} + C_{32}u_{2,2} + \left(C_{33} + \overset{\circ}{\sigma}_{33}\right)u_{3,3}.$$

It is easy to see that

$$\theta_{12} \neq \theta_{21}, \quad \theta_{23} \neq \theta_{32}, \quad \theta_{31} \neq \theta_{13}. \tag{27}$$

Further simplifications can be achieved introducing supplementary assumptions concerning the geometry of the body.

2. Representation of the incremental fields

The representation of elastic fields by complex potentials in the classical case of anisotropic elastic bodies was given by Leknitskii [8]. This representation was used, for instance, by Sih and Leibowitz [10] to analyze problems concerning the existence of a crack in an anisotropic elastic solid. The results obtained by Leknitskii were generalized for a prestressed material by Guz [5, 7], who also has analyzed the influence of the initial stresses on the behavior of a solid body containing cracks. We shall present Guz's representation of the incremental fields by complex potentials.

We assume that the orthotropic, initial deformed composite material is in a *plane state* relative to the x_1x_2 plane.

We have the following representation of the incremental fields by two arbitrary analytic complex potentials $\Phi_j = \Phi_j(z_j)$, $j = 1, 2$:

$$\theta_{22} = 2Re\left\{\Phi_1'(z_1) + \Phi_2'(z_2)\right\}, \tag{28}$$

$$\theta_{21} = -2Re\left\{a_1\mu_1\Phi_1'(z_1) + a_2\mu_2\Phi_2'(z_2)\right\}, \tag{29}$$

$$a_j = \frac{\omega_{2112}\omega_{1122}\mu_j^2 - \omega_{1111}\omega_{1212}}{B_j\mu_j^2}, \tag{30}$$

$$\theta_{12} = -2Re\left\{\mu_1\Phi_1'(z_1) + \mu_2\Phi_2'(z_2)\right\}, \tag{31}$$

$$\theta_{11} = 2Re\left\{a_1\mu_1^2\Phi_1'(z_1) + a_2\mu_2^2\Phi_2'(z_2)\right\}, \tag{32}$$

$$u_1 = 2Re\left\{b_1\Phi_1(z_1) + b_2\Phi_2(z_2)\right\}, \tag{33}$$

$$b_j = -\frac{\omega_{1122} + \omega_{1212}}{B_j}, \tag{34}$$

$$u_2 = 2Re\left\{c_1\Phi_1(z_1) + c_2\Phi_2(z_2)\right\}, \tag{35}$$

$$c_j = \frac{\omega_{2112}\mu_j^2 + \omega_{1111}}{B_j\mu_j}, \tag{36}$$

In these relations μ_k, $k = 1, 2$ are the roots of the incremental equation, and a_k, b_k, $k = 1, 2$ are coefficients depending on the instantaneous elasticities of the body and on the initial applied stress.

Now assume that orthotropic initial deformed composite material is in an *antiplane* state relative to the $x_1 x_2$ plane.

We have the following representation of the incremental fields corresponding to the antiplane state by a single complex potential $\Phi_3 = \Phi_3(z_3)$ depending on the complex variable $z_3 = x_1 + \mu_3 x_2$

$$\begin{aligned}
\theta_{23} &= -2Re\Phi_3'(z_3), \\
\theta_{13} &= 2Re\mu_3\Phi_3'(z_3),
\end{aligned}$$

(37)

$$u_3 = -2\omega_{2332}^{-1}Re\mu_3^{-1}\Phi_3(z_3).$$

(38)

This representation reduces to that given by Lekhnitski [7] if the initial applied deformations and stresses vanish.

We shall use Guz's results to study the incremental elastic state in an infinite, prestressed composite containing a crack.

3.   The opening, sliding and tearing modes

We assume that the whole space is occupied by an *initially deformed orthotropic material*. We suppose that the composite contains a crack of length $2a > 0$ situated on the x_1 axis and has infinite extent in the direction of the x_3 axis. As usual in continuum mechanics, we suppose that the crack is represented as a *cut* having two faces.

We assume that the initial deformed equilibrium configuration of the body is *homogeneous* and *locally stable*.

To ensure homogeneity of the initial deformed, prestressed equilibrium configuration, we must suppose that the initial applied stress $\overset{\circ}{\sigma}$ satisfies the following conditions

$$\overset{\circ}{\sigma}_{21} = \overset{\circ}{\sigma}_{22} = \overset{\circ}{\sigma}_{31} = 0.$$

(39)

First we formulate and solve the crack problem corresponding to *the first, opening mode*.

According to our assumptions, the incremental state of the body is a *plane state* relative to the plane $x_1 x_2$. The nominal stresses θ_{21} and θ_{22} must satisfy the following boundary conditions on the two faces of the crack (represented as a cut):

$$\begin{aligned}
\theta_{21}(x_1, 0^+) &= \theta_{21}(x_1, 0^-) = 0 \quad \text{for} \quad |x_1| < a, \\
\theta_{22}(x_1, 0^+) &= \theta_{22}(x_1, 0^-) = -g(x_1) \quad \text{for} \quad |x_1| < a,
\end{aligned}$$

(40)

where $g = g(x_1)$ is the given value of the incremental normal force,

$$\lim_{r\to\infty} \{u_\alpha(x_1, x_2), \theta_{\alpha\beta}(x_1, x_2)\} = 0,$$

$$r = \sqrt{x_1^2 + x_2^2} \quad \text{and} \quad \alpha, \beta = 1, 2. \tag{41}$$

On the two faces of the crack, the following conditions must be satisfied

$$a_1\mu_1\Psi_1^+(x_1) + a_2\mu_2\Psi_2^+(x_1) + \bar{a}_1\bar{\mu}_1\overline{\Psi}_1^+(x_1) + \bar{a}_2\bar{\mu}_2\overline{\Psi}_2^-(x_1) = 0,$$
$$a_1\mu_1\Psi_1^-(x_1) + a_2\mu_2\Psi_2^-(x_1) + \bar{a}_1\bar{\mu}_1\overline{\Psi}_1^+(x_1) + \bar{a}_2\bar{\mu}_2\overline{\Psi}_2^+(x_1) = 0, \tag{42}$$
$$|x_1| < a,$$

and

$$\Psi_1^+(x_1) + \Psi_2^+(x_1) + \overline{\Psi}_1^-(x_1) + \overline{\Psi}_2^-(x_1) = -g(x_1),$$
$$\Psi_1^-(x_1) + \Psi_2^-(x_1) + \overline{\Psi}_1^+(x_1) + \overline{\Psi}_2^+(x_1) = -g(x_1), \quad |x_1| < a. \tag{43}$$

Adding and subtracting Eqs. (42) and (43) we get

$$\left(\alpha\Psi_1(x_1) + \bar{\alpha}\overline{\Psi}_1(x_1)\right)^+ + \left(\alpha\Psi_1(x_1) + \bar{\alpha}\overline{\Psi}_1(x_1)\right)^- = -2g(x_1),$$
$$\left(\alpha\Psi_1(x_1) - \bar{\alpha}\overline{\Psi}_1(x_1)\right)^+ - \left(\alpha\Psi_1(x_1) - \bar{\alpha}\overline{\Psi}_1(x_1)\right)^- = 0, \quad |x_1| < a. \tag{44}$$

The first relation a nonhomogeneous Hilbert-Riemann problem. The general solution is

$$\alpha\Psi_1(z_1) + \bar{\alpha}\overline{\Psi}_1(z_1) = -\frac{X(z_1)}{\pi i} \int_{-a}^{a} \frac{g(t)dt}{X^+(t)(t-z)}, \tag{45}$$

where $\alpha = (a_2\mu_2 - a_1\mu_1)/a_2\mu_2$.

The second relation represents a homogeneous jump problem, thus we get

$$\alpha\Psi_1(z_1) - \bar{\alpha}\overline{\Psi}_1(z_1) = 0. \tag{46}$$

After elementary computations we can conclude that the complex potentials are given by the following relations

$$\Psi_1(z_1) = \Phi_1'(z_1) = -\frac{a_2\mu_2}{2\pi\Delta\sqrt{z_1^2 - a^2}} \int_{-a}^{a} \frac{g(t)\sqrt{a^2 - t^2}}{t - z_1} dt,$$

$$\Psi_2(z_2) = \Phi_2'(z_2) = \frac{a_1\mu_1}{2\pi\Delta\sqrt{z_2^2 - a^2}} \int_{-a}^{a} \frac{g(t)\sqrt{a^2 - t^2}}{t - z_2} dt, \tag{47}$$

where $\Delta = a_2\mu_2 - a_1\mu_1$.

We formulate and solve the crack problem corresponding to the *second, sliding mode*. More exactly, we assume that the incremental normal forces acting on the two faces of the crack vanish. Also, we suppose that on the two faces of the crack, antisymmetrically applied incremental tangential forces act in the direction of the x_1 axis.

The incremental state of the body is again a plane state relative to the plane $x_1 x_2$,

$$
\left(\alpha\Psi_1(x_1) + \overline{\alpha}\overline{\Psi}_1(x_1)\right)^+ + \left(\alpha\Psi_1(x_1) + \overline{\alpha}\overline{\Psi}_1(x_1)\right)^- = -2g(x_1),
$$
$$
\left(\alpha\Psi_1(x_1) - \overline{\alpha}\overline{\Psi}_1(x_1)\right)^+ - \left(\alpha\Psi_1(x_1) - \overline{\alpha}\overline{\Psi}_1(x_1)\right)^- = 0, \quad |x_1| < a.
$$
(48)

where $h = h(x_1)$ is the given incremental tangential force.

Using the same approach as in the Mode I, we get

$$
\Psi_1(z_1) = \Phi_1'(z_1) = -\frac{1}{2\pi\Delta\sqrt{z_2^2 - a^2}} \int_{-a}^{a} \frac{h(t)\sqrt{a^2 - t^2}}{t - z_1} dt,
$$
$$
\Psi_2(z_2) = \Phi_2'(z_2) = \frac{1}{2\pi\Delta\sqrt{z_1^2 - a^2}} \int_{-a}^{a} \frac{h(t)\sqrt{a^2 - t^2}}{t - z_2} dt.
$$
(49)

We formulate and solve the crack problem corresponding to *the third, tearing mode*.

On the two faces of the crack only tangential forces act, antisymmetrically distributed relative to the plane $x_2 = 0$ and having the direction of the x_3 axis.

The incremental state of the composite will be an antiplane state relative to the $x_1 x_2$ plane.

$$
\theta_{23}(x_1, 0^+) = \theta_{23}(x_1, 0^-) = -k(x_1) \quad \text{for} \quad |x_1| < a, \qquad (50)
$$

$k = k(x_1)$ is the given value of the applied incremental tangential surface force.

Using same approach as in Mode I, we get

$$
\Psi_3(z_3) = \Phi_3'(z_3) = \frac{1}{2\pi\Delta\sqrt{z_3^2 - a^2}} \int_{-a}^{a} \frac{h(t)\sqrt{a^2 - t^2}k(t)}{t - z_3} dt. \qquad (51)
$$

4.   Asymptotic behavior of the incremental fields

The incremental field distribution around the (right) tip can be obtained by letting

$$
x_1 = a + r\cos\varphi, \quad x_2 = r\sin\varphi \qquad (52)
$$

and by assuming that r is *small* in comparison with the half crack length a. The polar coordinate r and φ designate, respectively, the radial distance from the crack tip, and the angle between the radial line and the line extending the crack.

4.1   The first mode

In a small neighborhood of the crack tip, the Plemelj functions may be approximated by using

$$\sqrt{z_j^2 - a^2} = \sqrt{2ar}\chi_j(\varphi), \quad j = 1, 2, \tag{53}$$

where

$$\chi_j(\varphi) = \sqrt{\cos\varphi + \mu_j\sin\varphi}, \quad j = 1, 2. \tag{54}$$

The asymptotic values of the complex potentials are

$$\Psi_1(z_1) = \frac{K_I}{2\sqrt{2\pi r}}\frac{a_2\mu_2}{\Delta}\frac{1}{\chi_1(\varphi)},$$
$$\Psi_2(z_2) = \frac{K_I}{2\sqrt{2\pi r}}\frac{a_1\mu_1}{\Delta}\frac{1}{\chi_2(\varphi)}, \tag{55}$$

where

$$K_I = \frac{1}{\sqrt{\pi a}}\int_{-a}^{a} g(t)\sqrt{\frac{a+t}{a-t}}dt \tag{56}$$

is *the stress intensity factor*, corresponding to *the first mode*. This quantity has the *same* expression as in the classical theory of brittle fracture of elastic materials without initial stresses.

The asymptotic expressions of the incremental fields corresponding to the first mode are as follows:

$$\theta_{22} = \frac{K_I}{\sqrt{2\pi r}}Re\frac{1}{\Delta}\left\{\frac{a_2\mu_2}{\chi_1(\varphi)} - \frac{a_1\mu_1}{\chi_2(\varphi)}\right\}, \tag{57}$$

$$\theta_{21} = -\frac{K_I}{\sqrt{2\pi r}}Re\frac{a_1a_2\mu_1\mu_2}{\Delta}\left\{\frac{1}{\chi_1(\varphi)} - \frac{1}{\chi_2(\varphi)}\right\}, \tag{58}$$

$$\theta_{12} = -\frac{K_I}{\sqrt{2\pi r}}Re\frac{\mu_1\mu_2}{\Delta}\left\{\frac{a_2}{\chi_1(\varphi)} - \frac{a_1}{\chi_2(\varphi)}\right\}, \tag{59}$$

$$\theta_{11} = \frac{K_I}{\sqrt{2\pi r}}Re\frac{a_1a_2\mu_1\mu_2}{\Delta}\left\{\frac{\mu_1}{\chi_1(\varphi)} - \frac{\mu_2}{\chi_2(\varphi)}\right\}, \tag{60}$$

$$u_1 = 2\sqrt{\frac{r}{2\pi}}K_I Re\frac{1}{\Delta}\left\{b_1a_2\mu_2\chi_1(\varphi) - b_2a_1\mu_1\chi_2(\varphi)\right\}, \tag{61}$$

$$u_2 = 2\sqrt{\frac{r}{2\pi}}K_I Re\frac{1}{\Delta}\left\{c_1a_2\mu_2\chi_1(\varphi) - c_2a_1\mu_1\chi_2(\varphi)\right\}. \tag{62}$$

The incremental fields in a prestressed body have the *same* asymptotic behavior as the corresponding fields in an elastic body without initial stresses.

Particularly, since the incremental nominal stress behaves like $\frac{1}{\sqrt{r}}$ in the neighborhood of the crack tip, the total incremental elastic energy is *bounded* in any finite domain containing the crack. This fact is essential in *crack stability analysis* based on a Griffith's type criterion of crack propagation.

4.2 The second mode

The asymptotic values of the complex potentials are

$$
\begin{aligned}
\Psi_1(z_1) &= \frac{K_{II}}{2\sqrt{2\pi r}} \frac{1}{\Delta} \frac{1}{\chi_1(\varphi)}, \\
\Psi_2(z_2) &= -\frac{K_{II}}{2\sqrt{2\pi r}} \frac{1}{\Delta} \frac{1}{\chi_2(\varphi)},
\end{aligned}
\tag{63}
$$

where

$$
K_{II} = \frac{1}{\sqrt{\pi a}} \int_{-a}^{a} h(t) \sqrt{\frac{a+t}{a-t}} \, dt
\tag{64}
$$

is the *stress intensity factor*, corresponding to the second mode.

The asymptotic expression of the incremental fields corresponding to the second mode have the following expressions:

$$
\theta_{22} = \frac{K_{II}}{\sqrt{2\pi r}} Re \frac{1}{\Delta} \left\{ \frac{1}{\chi_1(\varphi)} - \frac{1}{\chi_2(\varphi)} \right\},
\tag{65}
$$

$$
\theta_{21} = -\frac{K_{II}}{\sqrt{2\pi r}} Re \frac{1}{\Delta} \left\{ \frac{a_1\mu_1}{\chi_1(\varphi)} - \frac{a_2\mu_2}{\chi_2(\varphi)} \right\},
\tag{66}
$$

$$
\theta_{12} = -\frac{K_{II}}{\sqrt{2\pi r}} Re \frac{1}{\Delta} \left\{ \frac{\mu_1}{\chi_1(\varphi)} - \frac{\mu_2}{\chi_2(\varphi)} \right\},
\tag{67}
$$

$$
\theta_{11} = \frac{K_{II}}{\sqrt{2\pi r}} Re \frac{1}{\Delta} \left\{ \frac{a_1\mu_1^2}{\chi_1(\varphi)} - \frac{a_2\mu_2^2}{\chi_2(\varphi)} \right\},
\tag{68}
$$

$$
u_1 = 2\sqrt{\frac{r}{2\pi}} K_{II} Re \frac{1}{\Delta} \{ b_1\chi_1(\varphi) - b_2\chi_2(\varphi) \},
\tag{69}
$$

$$
u_2 = 2\sqrt{\frac{r}{2\pi}} K_{II} Re \frac{1}{\Delta} \{ c_1\chi_1(\varphi) - c_2\chi_2(\varphi) \}.
\tag{70}
$$

The incremental fields in the prestressed body have the *same* asymptotic behavior as the corresponding fields, in an elastic body without initial stresses.

4.3   The third mode

The asymptotic expression of the involved complex potential:

$$\Psi_3(z_3) = -\frac{K_{\mathrm{III}}}{2\sqrt{2\pi r}}\frac{1}{\chi_3(\varphi)}, \tag{71}$$

where

$$\chi_3(\varphi) = \sqrt{\cos\varphi + \mu_3 \sin\varphi} \tag{72}$$

and

$$K_{\mathrm{III}} = \frac{1}{\sqrt{\pi a}}\int_{-a}^{a} k(t)\sqrt{\frac{a+t}{a-t}}\,\mathrm{d}t \tag{73}$$

is *the stress intensity factor* corresponding to *the third mode*.

The *asymptotic expressions of the incremental fields corresponding to the third* mode have the following expressions:

$$\theta_{22} = \frac{K_{\mathrm{III}}}{\sqrt{2\pi r}}Re\frac{1}{\chi_3(\varphi)}, \tag{74}$$

$$\theta_{12} = -\frac{K_{\mathrm{III}}}{\sqrt{2\pi r}}Re\frac{\mu_3}{\chi_3(\varphi)}, \tag{75}$$

$$u_3 = \frac{2}{\omega_{2332}}\sqrt{\frac{r}{2\pi}}K_{\mathrm{III}}Re\frac{\chi_3(\varphi)}{\mu_3}. \tag{76}$$

Until now, the quantity Δ is assumed to be non zero. However, if the material is initially deformed, then may be some *critical values* of the initial stress $\overset{\circ}{\sigma}_{11}$ for which Δ can be zero . If the initial applied stress $\overset{\circ}{\sigma}_{11}$ converges to this critical value, all incremental fields *increase unboundedly*.

When

$$\Delta \to 0 \tag{77}$$

resonance occurs.

We shall now analyze if the resonance can occur for a fiber-reinforced composite material.

The initial applied loading forces are in the fiber direction, parallel to the crack. We consider only the first and the second modes, since, for the third mode, resonance is not possible, as it can be seen by examining equations (74)–(76).

The critical value $\overset{\circ}{\sigma}_{11}^{cr}$ leading to resonance $\Delta = 0$ is

$$\overset{\circ}{\sigma}_{11}^{cr} = -C_{66}\left\{1 - \frac{C_{66}^2}{C_{11}C_{22}}\frac{C_{11}^2 C_{22}^2}{(C_{11}C_{22} - C_{12}^2)^2}\right\}. \tag{78}$$

This value is the same as the critical stress $\overset{\circ}{\sigma}_{11}^{\,cs}$ for which *surface instability* of the fiber-reinforced composite occurs [1]:

$$\overset{\circ}{\sigma}_{11}^{\,cr} = \overset{\circ}{\sigma}_{11}^{\,cs}. \tag{79}$$

In a fiber-reinforced composite, surface instability and resonance occur for the *same* critical value of the *compressive* force $\overset{\circ}{\sigma}_{11}$ acting in the direction of the reinforcing fibers. In a fiber-reinforced composite, material stress, having magnitude corresponding to $\overset{\circ}{\sigma}_{11}^{\,cr} = \overset{\circ}{\sigma}_{11}^{\,cs}$, produces *infinitesimal* deformations; hence, resonance can really occur! Concerning the case of surface instability, the actual possibility of resonance for a fiber-reinforced composite is a direct consequence of the existing *internal structure* of such materials! In a linearly elastic isotropic material, the resonance cannot occur, for infinitesimal deformations.

5.   Griffith's criterion and crack propagation

The mechanical and physical interpretations of the concept of stress intensity factor become clearer when the mathematical relationships between the rate of input work (or strain energy release rate) into the fracture process and the singular (asymptotic) elasticity solution in the neighborhood of the crack tip are established. Following Sih and Leibowitz [10], we suppose that a crack of length $2a$ in a plate of unit thickness is extended by the increment δa at both ends of the crack. This extension creates new surfaces of the crack with $4\delta a$ as the gain in area; the surface energy is increased by $4\gamma\delta a 4$, γ representing the specific surface energy of the body. $U(a)$ is the elastic energy of the body when the length of the crack is $2a$.

Griffith's criterion represents a necessary condition for the crack to propagate, i.e. the change in strain energy must satisfy the inequality

$$U(a) - U(a + \delta a) \geq 4\gamma\delta a. \tag{80}$$

We have

$$\delta U = U(a) - U(a + \delta a) = -\frac{\partial U}{\partial a}(a)\delta a. \tag{81}$$

The strain energy release rate $G(a)$ is defined by

$$G(a) = -\frac{1}{2}\frac{\partial U}{\partial a}(a). \tag{82}$$

From (80)−(82) we get

$$\delta U = U(a) - U(a + \delta a) = -2G(a)\delta a, \tag{83}$$

and Griffith's crack propagation or crack instability condition become

$$G(a) \geq 2. \tag{84}$$

The energy release rate $G(a)$ may by regarded as the force tending to open the crack. As we shall see, its evaluation requires *only* a knowledge of the incremental nominal stress and displacements near the crack tips. In the sequel, the ∂U and $G(a)$ values for the three basic modes of crack extension will be distinguished by the subscripts I, II and III, as we did for the case of the stress intensity factors.

5.1   The first opening mode

The strain energy released is the work done in this process by the incremental nominal stress $\theta_{22}(\delta a - t', 0)$ acting on the incremental displacement $u_2(t, 0^+)$, provided that δa is very small.

The variation δU_{I} of the elastic energy is due to the work done at *both* ends of the crack.

$$\delta U_{\mathrm{I}} = -2 \int_0^{\delta a} \theta_{22}(\delta a - t, 0) u_2(t, 0^+) \mathrm{d}t. \tag{85}$$

$$\delta U = U(a) - U(a + \delta a) = -2G(a)\delta a. \tag{86}$$

From the last equation we obtain the following relation:

$$G_{\mathrm{I}}(a)\delta = \int_0^{\delta a} \theta_{22}(\delta a - t, 0) u_2(t, 0^+) \mathrm{d}t. \tag{87}$$

The incremental nominal stress $\theta_{22}(\delta a - t, 0)$ has the following form:

$$\theta_{22}(\delta a - t, 0) = \frac{K_{\mathrm{I}}}{\sqrt{2\pi(\delta a - t)}}. \tag{88}$$

The incremental normal displacement $u_2(t, 0^+)$ of the upper face is

$$u_2(t, 0^+) = 2K_{\mathrm{I}}\sqrt{\frac{t}{2\pi}} Re\left(i\frac{c_1 a_2 \mu_2 - c_2 a_1 \mu_1}{\Delta}\right). \tag{89}$$

We now use the relation

$$\frac{c_1 a_2 \mu_2 - c_2 a_1 \mu_1}{\Delta} = \mu_1 \mu_2 \frac{l}{f}, \tag{90}$$

and get

$$u_2(t, 0^+) = 2K_{\mathrm{I}}\sqrt{\frac{t}{2\pi}} Re\left(i\mu_1\mu_2\frac{l}{f}\right), \tag{91}$$

with

$$\begin{aligned} f = {} & \omega_{1122}\omega_{2112}[\omega_{1111}\omega_{2222} - \omega_{1122}(\omega_{1122} + \omega_{1212}) \\ & - \omega_{2222}\omega_{2112}\mu_1\mu_2]\mu_1^2\mu_2^2 + \omega_{1111}\omega_{1212}\{\omega_{1111}\omega_{1221} \\ & + [\omega_{1212}(\omega_{1122} + \omega_{1212}) - \omega_{2112}\omega_{1221}]\mu_1\mu_2\}. \end{aligned} \tag{92}$$

$$l = \omega_{1111}\omega_{2112}(\omega_{1122} + \omega_{1212})(\mu_1 + \mu_2) \qquad (93)$$

and

$$\hat{l} = il. \qquad (94)$$

We get

$$u_2(t, 0^+) = 2K_I\mu_1\mu_2\frac{\hat{l}}{f}\sqrt{\frac{t}{2\pi}}. \qquad (95)$$

Consequently, we have

$$G_I(a)\delta a = \frac{K_I^2}{\pi}\frac{\mu_1\mu_2\hat{l}}{f}\int_0^{\delta a}\sqrt{\frac{t}{\delta a - t}}dt, \qquad (96)$$

or equivalent by

$$G_I(a) = \frac{K_I^2}{2}\frac{\mu_1\mu_2\hat{l}}{f}. \qquad (97)$$

Using a similar approach, we find

$$G_{II}(a)\delta a = \int_0^{\delta a}\theta_{21}(\delta a - t, 0)\, u_1(t, 0^+)\mathrm{d}t, \qquad (98)$$

$$\theta_{21}(\delta a - t, 0) = \frac{K_{II}}{\sqrt{2\pi(\delta a - t)}}, \qquad (99)$$

$$u_1(t, 0^+) = 2K_{II}\sqrt{\frac{t}{2\pi}}Re\left(i\frac{b_1 - b_2}{\Delta}\right), \qquad (100)$$

$$G_{II}(a) = \frac{K_{II}^2}{\pi}\frac{\mu_1^2\mu_2^2\hat{m}}{f}, \qquad (101)$$

where

$$\hat{m} = im, \qquad (102)$$

$$m = (\omega_{1122} + \omega_{1212})\omega_{2222}\omega_{2112}(\mu_1 + \mu_2), \qquad (103)$$

and

$$G_{III}(a) = \frac{K_{III}^2}{2\sqrt{\omega_{2332}\omega_{1331}}}. \qquad (104)$$

The *total* strain energy release rate $G(a)$ is the sum of the release rates corresponding to the three basic modes (97), (98) and (104):

$$G(a) = \frac{1}{2}K_I^2\mu_1\mu_2\hat{l}f^{-1} + \frac{1}{2}K_{II}^2\mu_1^2\mu_2^2\hat{m}f^{-1} + \frac{1}{2}K_{III}^2(\omega_{2332}\omega_{1331})^{-\frac{1}{2}}. \qquad (105)$$

Crack instability occurs and *crack propagation starts* if the applied incremental forces satisfy the *propagation condition*:

$$K_I^2\mu_1\mu_2\hat{l}f^{-1} + K_{II}^2\mu_1^2\mu_2^2\hat{m}f^{-1} + K_{III}^2(\omega_{2332}\omega_{1331})^{-\frac{1}{2}} = 4\gamma. \qquad (106)$$

We recall that the stress concentration factors K_I, K_{II}, K_{III} depend on the crack length and the normal and tangential incremental forces applied on the two faces of the crack. The roots μ_1, μ_2 and the coefficients f, $\hat{l}$, $\widehat{m}$, ω_{2332}, ω_{1331} depend on the elastic properties of the material and on the initial applied stresses. In a prestressed material, the specific surface energy of the material $\gamma > 0$ can depend also on the initial applied stresses. This dependence is not known at present. In the following, we *assume* that the dependence of the specific surface energy $\gamma > 0$ on the initial applied stresses is *negligible*, and γ depends only on the material properties. Clearly, this is a relatively strong assumption, and its validity must be checked by further research.

To simplify the analysis of the consequences of the Griffith's propagation criterion, we suppose constant incremental forces act on the crack faces corresponding to the three modes

$$
\begin{aligned}
g(x_1) &= p = const. > 0, \\
h(x_1) &= \tau = const. > 0, \\
k(x_1) &= \kappa = const. > 0 \quad \text{for} \quad |x_1| < a.
\end{aligned}
\tag{107}
$$

The stress concentration factors are

$$
\begin{aligned}
K_I &= \sqrt{\pi a}\, p, \\
K_{II} &= \sqrt{\pi a}\, \tau, \\
K_{III} &= \sqrt{\pi a}\, \kappa.
\end{aligned}
\tag{108}
$$

Griffith's propagation criterion become:

$$
p^2 \mu_1 \mu_2 \hat{l} f^{-1} + \tau^2 \mu_1^2 \mu_2^2 \widehat{m} f^{-1} + \kappa^2 (\omega_{2332}\omega_{1331})^{-\frac{1}{2}} = 4\gamma(\pi a)^{-1}.
\tag{109}
$$

We assume that only $\overset{\circ}{\sigma}_{11}$ is nonvanishing; i.e. we assume that

$$
\overset{\circ}{\sigma}_{22} = \overset{\circ}{\sigma}_{33} = 0.
\tag{110}
$$

We assume that the stress-free reference configuration of the material is locally stable and its initial deformed equilibrium configuration is internally (structurally) stable.

Let us assume that 4 the applied incremental tangential forces are vanishing; i.e.

$$
\tau = \kappa = 0.
\tag{111}
$$

In this case, Griffith's propagation criterion (109) becomes

$$
p^2 = \frac{4\gamma f}{\pi a \mu_1 \mu_2 \hat{l}}.
\tag{112}
$$

The critical value p_c for which the crack propagation starts in the first mode is

$$p_c^2 = p_c^2 \left(\overset{\circ}{\sigma}_{11} \right) = \frac{4\gamma f}{\sqrt{2\pi a} \left(C_{11} + \overset{\circ}{\sigma}_{11} \right) C_{66} \left(C_{12} + C_{66} \right) \sqrt{B}\sqrt{A + \sqrt{B}}}. \tag{113}$$

From (113) implies

$$p_c(\overset{\circ}{\sigma}_{11}) \to 0 \quad \text{when} \quad f\left(\overset{\circ}{\sigma}_{11} \right) \to 0. \tag{114}$$

We denote by $\overset{\circ}{\sigma}_{11}^{cp}$ the critical value of the initial applied stress $\overset{\circ}{\sigma}_{11}$ for which

$$f\left(\overset{\circ}{\sigma}_{11}^{cp} \right) = 0. \tag{115}$$

If the initial applied stress $\overset{\circ}{\sigma}_{11}$ reaches its critical value $\overset{\circ}{\sigma}_{11}^{cp}$, the crack becomes *completely unstable* and its propagation can start *without* any incremental normal force applied at the two faces of the crack!

Moreover we have

$$\overset{\circ}{\sigma}_{11}^{cp} = \overset{\circ}{\sigma}_{11}^{cr} = \overset{\circ}{\sigma}_{11}^{cs}, \tag{116}$$

i.e. complete instability of the crack, resonance and surface instability of a fiber-reinforced composite material occur for the same critical value of the initial applied compressive force, acting in the direction of the reinforcing fibers.

Let us assume now that

$$p = \kappa = 0. \tag{117}$$

In this case

$$\tau^2 = \frac{4\gamma f}{\pi a \mu_1^2 \mu_2^2 \widehat{m}}. \tag{118}$$

The critical value τ_c for which the crack propagation starts in the second mode is:

$$\tau_c^2 = \tau_c^2 \left(\overset{\circ}{\sigma}_{11} \right) = \frac{4\gamma f}{\sqrt{2\pi a} C_{22} C_{66} \left(C_{12} + C_{66} \right) \sqrt{B}\sqrt{A + \sqrt{B}}}. \tag{119}$$

Again can see that

$$\tau_c \left(\overset{\circ}{\sigma}_{11} \right) \to 0 \quad \text{if} \quad f\left(\overset{\circ}{\sigma}_{11} \right) \to 0. \tag{120}$$

Hence, when the initial applied compressive stress $\overset{\circ}{\sigma}_{11}$ reaches its critical value

$$\overset{\circ}{\sigma}_{11}^{cp} = \overset{\circ}{\sigma}_{11}^{cr} = \overset{\circ}{\sigma}_{11}^{cs} = 0, \tag{121}$$

i.e. the crack becomes completely unstable and its propagation can start with-out any incremental tangential force applied at the two faces of the crack and in the direction of the crack.

Now assume that only the third mode is non zero:

$$p = \tau = 0. \tag{122}$$

The propagation condition becomes

$$\kappa^2 = \frac{4\gamma\sqrt{\omega_{1331}\omega_{2332}}}{\pi a}. \tag{123}$$

Taking this relation into account and denoting by $\kappa_c = \kappa_c(\overset{\circ}{\sigma}_{11})$ the critical value of the applied tangential stress, we find

$$\kappa_c^2 = \kappa_c^2\left(\overset{\circ}{\sigma}_{11}\right) = \frac{4\gamma}{\pi a}\sqrt{C_{44}\left(C_{55} + \overset{\circ}{\sigma}_{11}\right)}. \tag{124}$$

We denote the critical value corresponding to $\overset{\circ}{\sigma}_{11} = 0$ by $\hat{\kappa}_c$; i.e.

$$\hat{\kappa}_c^2 = \kappa_c^2(0) = \frac{4\gamma}{\pi a}\sqrt{C_{44}C_{55}}. \tag{125}$$

We have

$$\kappa_c^2\left(\overset{\circ}{\sigma}_{11}\right) = \hat{\kappa}_c^2\sqrt{\frac{C_{55} + \overset{\circ}{\sigma}_{11}}{C_{55}}}. \tag{126}$$

Hence

$$\begin{aligned}\kappa\left(\overset{\circ}{\sigma}_{11}\right) > \hat{\kappa}_c \quad \text{if} \quad \overset{\circ}{\sigma}_{11} > 0, \\ \kappa\left(\overset{\circ}{\sigma}_{11}\right) < \hat{\kappa}_c \quad \text{if} \quad \overset{\circ}{\sigma}_{11} < 0.\end{aligned} \tag{127}$$

Consequently, an initial applied extensional force $\overset{\circ}{\sigma}_{11} > 0$ acting in the direction of the reinforcing fibers *improves* the crack stability; an initial applied compressive force $\overset{\circ}{\sigma}_{11} < 0$ acting in the fiber direction *diminishes* the stability of the crack.

We have

$$\kappa_c\left(\overset{\circ}{\sigma}_{11}\right) \to 0 \quad \text{if} \quad \overset{\circ}{\sigma}_{11} \to \overset{\circ}{\sigma}_{11}^{cp} = -C_{55}. \tag{128}$$

When the initial applied compressive stress $\overset{\circ}{\sigma}_{11}$ reaches the *critical value*,

$$\overset{\circ}{\sigma}_{11}^{cp} = -C_{55}, \tag{129}$$

the crack becomes completely unstable and its propagation in the third mode can start *without* any incremental tangential force applied to the two faces of the crack.

We recall that, for a fiber-reinforced composite, we have

$$C_{55} = G_{13} \quad \text{and} \quad G_{13} \ll E_1. \tag{130}$$

Hence, using the last two equations, we have

$$\overset{\circ}{\sigma}{}_{11}^{cp} = -G_{13} \quad \text{and} \quad \left| \overset{\circ}{\sigma}{}_{11}^{cp} \right| \ll E_1. \tag{131}$$

Complete instability of the crack can occur in a fiber-reinforced composite material if the condition (131) is fulfilled.

The *simultaneous* appearance of internal instability and complete instability of the crack are direct consequences of the internal structure of a fiber-reinforced composite. Neither complete crack instability nor internal instability can occur for an isotropic material in the framework of linear elasticity.

The analysis reveals the way in which a phenomenological continuum theory can take into account and predict macroscopic effects from the internal structure of a composite material.

We see that due to its internal structure, a dangerous situation can occur in a fiber-reinforced composite material, if the initial applied forces are not adequately limited. The three-dimensional linearized theory is able to reveal the critical situations and can be used to avoid the occurrence of dangerous situations.

Acknowledgments

The author gratefully acknowledge the support provided to Prof. Tomasz Sadowski – Coordinator of the European Project *Marie Curie Host Fellowships for Transfer of Knowledge (TOK)* – "Modern Composite Materials Applied in Aerospace, Civil and Mechanical Engineering: Theoretical Modelling and Experimental Verification".

References

[1] CRISTESCU N., CRACIUN E., SOÓS E. (2003): *Mechanics of Elastic Composites*. CRC Press, Chapmann & Hall.

[2] ERINGEN A., MAUGIN G. (1990): *Electrodynamics of Continua, vol. I, Foundations and Solid Media*. Springer, New York.

[3] GUZ A. (1971): *Stability of Three-Dimensional Deformable Bodies*. Naukova Dumka, Kiev, in Russian.

[4] GUZ A. (1983): *Mechanics of Brittle Fracture of Materials, with Initial Stresses*. Naukova Dumka, Kiev, in Russian.

[5] GUZ A. (1983): *Mechanics of Brittle Fracture of Prestressed Materials*. Visha Schola, Kiev, in Russian.

[6] GUZ A. (1986): *Fundamentals of Three-Dimensional Theory of Stability of Deformable Bodies*. Visha Schola, Kiev, in Russian.

[7] GUZ A. (1991): Brittle fracture of materials with initial stress. In: A. Guz (red.) *Non-Classical Problems of Fracture Mechanics*, tom II, Naukova Dumka, Kiev, in Russian, 27–74.

[8] LEKHNITSKI S. (1963): *Theory of Elasticity of Aniosotropic Elastic Body*. Holden Day, San Francisco.

[9] OGDEN R. (1984): *Non-Linear Elastic Deformations*. Wiley, New York.

[10] SIH G., LEIBOWITZ H. (1968): Mathematical theories of brittle fracture. W: H. Lebowitz (red.) *Fracture – An Advanced Treatise*, Vol. II. Mathematical Fundamentals, str. 68–191, Academic Press, New York.

ENERGY CRITERIA FOR CRACK PROPAGATION IN PRE-STRESSED ELASTIC COMPOSITES

Eduard Marius Craciun

Faculty of Mathematics and Informatics
Ovidius University Constanta
blvd. Mamaia 124, cp 900527
Romania
mcraciun@univ-ovidius.ro

Abstract   We study the interaction of two unequal collinear cracks in a pre-stressed fiber reinforced elastic composite in Modes I and II of classical fracture. Using the theory of Riemann – Hilbert problem, Plemelj's function and the theory of Cauchy's integral we decide which tip of the crack will start to propagate first. We generalize Sih's fracture criterion for Modes I, II and we determine the direction of propagation for two transversally isotropic materials, graphite/epoxy and aramid/epoxy. The resonance phenomenon is studied in the case of unequal collinear cracks.

Keywords: collinear cracks, pre-stressed elastic composite, crack interaction, Sih's fracture criterion

1.   Interaction of two unequal cracks in a pre-stressed fiber reinforced composite

We consider a pre-stressed, orthotropic, linear, elastic material representing a fiber reinforced composite. We take as coordinate planes the symmetry planes of the material, the x_1-axis being parallel to the fiber. We assume that the admissible incremental equilibrium states of the body are plane strain states relative to the $x_1 x_2$-plane. As was shown by Guz (see [5–8]), in these assumed circumstances the incremental equilibrium states of the material can be represented by two complex potentials defined in two complex planes. In Sect. 1 we give Guz's representation formula.

R. de Borst and T. Sadowski (eds.) *Lecture Notes on Composite Materials – Current Topics and Achievements*

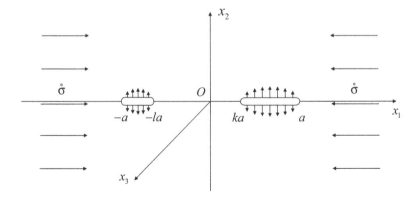

Figure 1. Two unequal cracks

We suppose that the material is unbounded and contains two collinear and unequal cracks situated in the same plane, parallel to the reinforcing fibers and to the initial applied stresses (see Fig. 1). We assume that the upper and lower faces of the cracks are acted on by symmetrically distributed normal incremental stresses. Our first aim is to determine the incremental elastic state produced in the body. To do this we use Guz's representation theorem. Using the theory of the Riemann-Hilbert problem we give the solution of our mathematical problem, assuming that the applied incremental stresses have a given constant value. Our second aim is to determine the critical values of the given incremental stress for which the tips of the cracks start to propagate. To do this we determine the singular parts of the elastic state near the crack tips, and we use Irwin's relation giving the energy release rate, as well as Griffith's energy criterion.

Another aim is to study the interaction of the two cracks. For an isotropic material, without initial applied stresses, we obtain Willmore's classical results for the critical stresses producing crack propagation. We also study the interaction of cracks having different lengths. The results show that this interaction is strong only if the distance between the cracks is much smaller than their length. In such a situation, the inner tips start to propagate first and the cracks tend to unify. If this distance is much greater than the length of the cracks, the interaction is weak, and the tips of the longest crack start to propagate first.

1.1   Unequal cracks in a pre-stressed fiber reinforced composite. The first fracture mode

We consider a pre-stressed, orthotropic, linear, elastic material representing a fiber reinforced composite. We take as coordinate planes the symmetry planes of the material, the x_1-axis being parallel to the fiber. We assume that the admissible incremental equilibrium states of he body are plane strain states relative to the $x_1 x_2$-plane. As it was shown by Guz (see [6, 7]), in these circumstances, the incremental elastic state of the material can be expressed by two analytic complex potentials $\Phi_j(z_j)$ defined in two complex planes z_j, $j = 1, 2$. We denote the components of the incremental displacement field by u_1, u_2 and the involved components of the incremental nominal stress by θ_{11}, θ_{12}, θ_{21}, θ_{22}. We present Guz's representation formulae in the weakly modified form due to Soos (see [18]). We have

$$u_1 = 2Re\{b_1\Phi_1(z_1) + b_2\Phi_2(z_2)\}, \quad u_2 = 2Re\{c_1\Phi_1(z_1) + c_2\Phi_2(z_2)\}, \quad (1)$$

$$
\begin{aligned}
\theta_{11} &= 2Re\{a_1\mu_1^2\Psi_1(z_1) + a_2\mu_2^2\Psi_2(z_2)\}, \\
\theta_{12} &= -2Re\{\mu_1\Psi_1(z_1) + \mu_2\Psi_2(z_2)\},
\end{aligned}
\quad (2)
$$

$$
\begin{aligned}
\theta_{21} &= -2Re\{a_1\mu_1\Psi_1(z_1) + a_2\mu_2\Psi_2(z_2)\}, \\
\theta_{22} &= 2Re\{\Psi_1(z_1) + \Psi_2(z_2)\}.
\end{aligned}
\quad (3)
$$

In this relations

$$\Psi_j(z_j) = \Phi_j'(z_j) = \frac{d\Phi_j(z_j)}{dz_j}, \quad j = 1, 2 \quad (4)$$

and

$$z_j = x_1 + \mu_j x_2, \quad j = 1, 2. \quad (5)$$

The parameters μ_j are the roots of the algebraic equation

$$\mu^4 + 2A\mu^2 + B = 0 \quad (6)$$

with

$$
\begin{aligned}
A &= \frac{\omega_{1111}\omega_{2222} + \omega_{1221}\omega_{2112} - (\omega_{1122} + \omega_{1212})^2}{2\omega_{2222}\omega_{2112}}, \\
B &= \frac{\omega_{1111}\omega_{1221}}{\omega_{2222}\omega_{2112}}.
\end{aligned}
\quad (7)
$$

The parameters a_j, b_j, c_j, $j = 1, 2$ have the following expressions:

$$a_j = (\omega_{2112}\omega_{1122}\mu_j^2 - \omega_{1111}\omega_{1212})/(B_j\mu_j^2), \quad (8)$$

$$b_j = -(\omega_{1122} + \omega_{1212})/B_j, c_j = (\omega_{2112}\mu_j^2 + \omega_{1111})/(B_j\mu_j), \quad (9)$$

with

$$B_j = \omega_{2222}\omega_{2112}\mu_j^2 + \omega_{1111}\omega_{2222} - \omega_{1122}(\omega_{1122} + \omega_{1212}). \quad (10)$$

The instantaneous elastic moduli ω_{klmn}, $k, l, m, n = 1, 2$, involved in the above relations can be expressed in terms of the elastic coefficients C_{11}, C_{12},

C_{22}, C_{66} of the material and the initial applied stress $\overset{\circ}{\sigma}$ by the following relations:

$$\omega_{1111} = C_{11} + \overset{\circ}{\sigma}, \quad \omega_{2222} = C_{22}, \quad \omega_{1122} = C_{12}.$$
$$\omega_{1212} = C_{66}, \omega_{1221} = C_{66} + \overset{\circ}{\sigma}, \quad \omega_{2112} = C_{66}. \tag{11}$$

In their turn the elastic coefficients can be expressed using the engineering constants of the composite; we have

$$C_{11} = \frac{1 - \nu_{23}\nu_{32}}{E_2 E_3 D}, \quad C_{22} = \frac{1 - \nu_{13}\nu_{31}}{E_1 E_3 D},$$
$$C_{12} = \frac{\nu_{12} + \nu_{32}\nu_{13}}{E_1 E_3 H}, \quad C_{66} = G_{12}, \tag{12}$$

with

$$D = \frac{\left(1 - \nu_{12}\nu_{21} - \nu_{23}\nu_{32} - \nu_{31}\nu_{13} - \nu_{21}\nu_{32}\nu_{13} - \nu_{12}\nu_{23}\nu_{31}\right)}{(E_1 E_2 E_3)}. \tag{13}$$

In this relations E_1, E_2, E_3 are Young's moduli in the corresponding symmetry directions of the material; $\nu_{12}, \ldots, \nu_{32}$ are Poisson's ratios and G_{12} is the shear modulus in the symmetry plane $x_1 x_2$. By $\overset{\circ}{\sigma}$ we have denoted the initial applied stress acting in the fiber direction.

It can be shown that if the initial deformed equilibrium configuration of the body is locally stable; the roots of the algebraic Eq. (6) are complex numbers, having nonvanishing imaginary parts. Also, for an orthotropic composite μ_1 and μ_2 are not equal, i.e.,

$$\mu_1 \neq \mu_2. \tag{14}$$

Ending this section we observe that, for our fiber reinforced composite material, the Young's modulus E_1 is much greater than the Young's modulus E_2 and the shear modulus G_{12}; i.e.,

$$E_2 \ll E_1, \ G_{12} \ll E_1. \tag{15}$$

More details concerning these elements of three-dimensional linearized theory of elastic bodies can be found in Guz's fundamental monograph (Guz, [6]).

We shall use Guz's complex representation of the incremental elastic state to study the interaction between two cracks having different lengths in a prestressed fiber reinforced composite material.

We assume an unbounded material. We suppose that the material contains two collinear and unequal cracks situated in the same plane, parallel to the reinforcing fiber, and with initial applied stress, as in Fig. 1. The plane containing the cracks is taken as the $x_1 x_3$, the axis x_2-axis being perpendicular to the cracks faces.

We suppose that on the upper and lower faces of the cracks there are given normal incremental stresses, symmetrically distributed relative to the plane containing the cracks. The incremental elastic state is a plane deformation state, relative to the $x_1 x_2$-plane. Hence we can use Guz's complex representation to study our problem, corresponding to the first mode in classical fracture mechanics.

We denote by $\mathcal{L}$ the cut corresponding to the segments $(-a, -la)$ and (ka, a) as in Fig. 1. Here $a > 0$ is a given positive constant, and l, k are given positive numbers satisfying the restriction $0 < l, k < 1$. Obviously, $-a$ and $-la$ are the coordinates of the tips of the left crack, and ka and a of the right.

We denote by

$$p = p(t), \ t \in (-a, -la) \cup (ka, a), \tag{16}$$

the given incremental normal stress.

The components θ_{21} and θ_{22} of the incremental nominal stress must satisfy the following boundary conditions

$$\theta_{21}(t, 0^+) = \theta_{21}(t, 0^-) = 0,$$
$$\theta_{22}(t, 0^+) = \theta_{22}(t, 0^-) = -p(t), \tag{17}$$

for $t \in (-a, -la) \cup (ka, a)$.

Also, at large distances from the cracks, the incremental displacements and nominal stresses must vanish; i.e.,

$$\lim_{r \to \infty} \{u_\alpha(x_1, x_2), \theta_{\alpha\beta}(x_1, x_2)\} = 0, \ r = \sqrt{x_1^2 + x_2^2}, \quad \alpha, \beta = 1, 2. \tag{18}$$

Consequently, from (1)–(3) we can conclude that Guz's complex potential must satisfy the conditions

$$\lim_{z_j \to \infty} \{\Phi_j(z_j), \Psi_j(z_j)\} = 0, \quad j = 1, 2. \tag{19}$$

From Guz's representation formula (1) and from the boundary conditions (16) it follows that the complex potentials $\Psi_j(z_j)$ must satisfy the following conditions

$$a_1 \mu_1 \Psi_1^+ + a_2 \mu_2 \Psi_2^+ + \overline{a_1 \mu_1} \bar{\Psi}_1^- + \overline{a_2 \mu_2} \bar{\Psi}_2^- = 0,$$
$$a_1 \mu_1 \Psi_1^- + a_2 \mu_2 \Psi_2^- + \overline{a_1 \mu_1} \bar{\Psi}_1^+ + \overline{a_2 \mu_2} \bar{\Psi}_2^+ = 0, \tag{20}$$

for $t \in (-a, -la) \cup (ka, a)$.

Adding and subtracting these relations leads to

$$(a_1 \mu_1 \Psi_1 + a_2 \mu_2 \Psi_2 + \overline{a_1 \mu_1} \bar{\Psi}_1 + \overline{a_2 \mu_2} \bar{\Psi}_2)^+$$
$$+ (a_1 \mu_1 \Psi_1 + a_2 \mu_2 \Psi_2 + \overline{a_1 \mu_1} \bar{\Psi}_1 + \overline{a_2 \mu_2} \bar{\Psi}_2)^- = 0, \tag{21}$$

$$(a_1\mu_1\Psi_1 + a_2\mu_2\Psi_2 - \overline{a_1\mu_1}\bar{\Psi}_1 - \overline{a_2\mu_2}\bar{\Psi}_2)^+$$
$$- (a_1\mu_1\Psi_1 + a_2\mu_2\Psi_2 - \overline{a_1\mu_1}\bar{\Psi}_1 - \overline{a_2\mu_2}\bar{\Psi}_2)^- = 0. \tag{22}$$

Using well known proprieties of Cauchy's integral ([12], Chap. 4) and taking into account the conditions (19) at a large distance, from (22) we can conclude that the complex potentials must satisfy the following relations in the whole complex plane z:

$$a_1\mu_1\Psi_1(z) + a_2\mu_2\Psi_2(z) - \overline{a_1\mu_1}\bar{\Psi}_1(z) - \overline{a_2\mu_2}\bar{\Psi}_2(z) = 0. \tag{23}$$

The restriction (21) is equivalent to a homogeneous Riemann-Hilbert problem (see [12], Chap. 6) and its general solution is given by the relation

$$a_1\mu_1\Psi_1(z) + a_2\mu_2\Psi_2(z) + \overline{a_1\mu_1}\bar{\Psi}_1(z) + \overline{a_2\mu_2}\bar{\Psi}_2(z) = Q_0(z)X(z) \tag{24}$$

where

$$X(z) = \frac{1}{\sqrt{(z^2 - a^2)(z + la)(z - ka)}} \tag{25}$$

is the Plemelj function corresponding to the cut $\mathcal{L}$, and $Q_0(z)$ is an arbitrary polynomial having first degree and complex coefficients. It what follows we assume that

$$Q_0(z) \equiv 0. \tag{26}$$

As we shall see later, our choice is justified, since assuming (26) we shall be able to satisfy all boundary conditions. Hence the uniqueness theorem says us that no other choice is possible. Using (26) from (24) we get the second restriction which must be satisfied by the complex potentials in the whole complex plane z

$$a_1\mu_1\Psi_1(z) + a_2\mu_2\Psi_2(z) + \overline{a_1\mu_1}\bar{\Psi}_1(z) + \overline{a_2\mu_2}\bar{\Psi}_2(z) = 0. \tag{27}$$

Equations (23) and (27) lead to

$$a_1\mu_1\Psi_1(z) + a_2\mu_2\Psi_2(z) = 0. \tag{28}$$

Hence

$$\Psi_2(z) = -\frac{a_1\mu_1}{a_2\mu_2}\Psi_1(z). \tag{29}$$

Using this result, and taking into account the boundary conditions (17) we can conclude that the complex potential $\Psi_1(z_1)$ must satisfy the following conditions

$$\rho\Psi_1^+(t) + \bar{\rho}\bar{\Psi}_1^-(t) = -p(t), \quad \rho\Psi_1^-(t) + \bar{\rho}\bar{\Psi}_1^+(t) = p(t) \tag{30}$$

for $t \in (-a, -la) \cup (ka, a)$.

In these relations we have used the following notation:

$$\rho = \frac{\Delta}{a_2 \mu_2} \tag{31}$$

and

$$\Delta = a_2 \mu_2 - a_1 \mu_1. \tag{32}$$

Using the same procedure as before, from (30) we get the following equivalent conditions

$$(\rho \Psi_1(t) + \bar{\rho}\bar{\Psi}_1(t))^+ + (\rho \Psi_1(t) + \bar{\rho}\bar{\Psi}_1(t))^- = -2p(t), \tag{33}$$

$$(\rho \Psi_1(t) - \bar{\rho}\bar{\Psi}_1(t))^+ - (\rho \Psi_1(t) - \bar{\rho}\bar{\Psi}_1(t))^- = 0, \tag{34}$$

for $t \in (-a, -la) \cup (ka, a)$.

From the boundary condition (34), we can conclude, as before, that the potential Ψ_1 must satisfy the following relation in the whole complex plane z:

$$\rho \Psi_1(z) - \bar{\rho}\bar{\Psi}_1(z) = 0. \tag{35}$$

The restriction (33) is equivalent to a nonhomogeneous Riemann-Hilbert problem. Its solution can be obtained using the Plemelj function $X(z)$ introduced by the relation (25). We shall use that branch of this multivalued function which satisfies the conditions

$$\lim_{z \to \infty} (1/X(z)) = +\infty. \tag{36}$$

As it is easy to see, the upper and lower limits of this branch are given by the following relations

$$\begin{aligned}
X^+(t) &= \frac{1}{ia(t)}, & X^-(t) &= -\frac{1}{ia(t)} & \text{for } ka < t < a, \\
X^+(t) &= -\frac{1}{ia(t)}, & X^-(t) &= \frac{1}{ia(t)} & \text{for } -a < t < -la,
\end{aligned} \tag{37}$$

where

$$a(t) = \sqrt{(a^2 - t^2)(t + la)(t - ka)} > 0 \quad \text{for} \quad t \in (-a, -la) \cup (ka, a). \tag{38}$$

Under these conditions, the general solution of the nonhomogeneous Riemann-Hilbert problem is given by

$$\rho \Psi_1(z) + \bar{\rho}\bar{\Psi}_1(z) = -\frac{X(z)}{\pi i} \int_L \frac{p(t)dt}{X^+(t)(t - z)} + P(z)X(z). \tag{39}$$

In this equation $\mathcal{L} = (-a, -la) \cup (ka, a)$ and $P(z)$ is an arbitrary polynomial having complex coefficients. Taking into account the restriction (19) at large distance we can conclude that $P(z)$ must be a first degree polynomial; i.e.,

$$P(z) = C_1 z + C_2, \tag{40}$$

where C_1 and C_2 are arbitrary complex numbers. Later we shall see that these constants can be uniquely determined by imposing the uniformity of the incremental displacement field.

Now, from (29), (31), (35) and (39) we get the following expressions for the complex potentials:

$$\Psi_1(z_1) = -\frac{a_2\mu_2}{\Delta} \frac{X(z_1)}{2\pi i} \int_{\mathcal{L}} \frac{p(t)dt}{X^+(t)(t - z_1)} + \frac{a_2\mu_2}{2\Delta} P(z_1)X(z_1) \tag{41}$$

and

$$\Psi_2(z_2) = \frac{a_1\mu_1}{\Delta} \frac{X(z_2)}{2\pi i} \int_{\mathcal{L}} \frac{p(t)dt}{X^+(t)(t - z_2)} - \frac{a_1\mu_1}{2\Delta} P(z_2)X(z_2). \tag{42}$$

where Δ is defined by (32).

In all what follows we assume that the given incremental stresses acting on the cracks faces have a constant value; i.e.,

$$p(t) = p = const. > 0 \quad \text{for} \quad t \in (-a, -la) \cup (ka, a). \tag{43}$$

In this case the integrals involved in (41) and (42) can be calculated using known results of the theory of complex functions (see [12]). Finally for the complex potentials we get the expressions

$$\Psi_1(z_1) = \frac{pa_2\mu_2}{2\Delta} \left(\frac{z_1^2 - \alpha z_1 - \beta^2}{\sqrt{(z_1^2 - a^2)(z_1 + la)(z_1 - ka)}} - 1 \right)$$
$$+ \frac{a_2\mu_2}{2\Delta} \frac{P(z_1)}{\sqrt{(z_1^2 - a^2)(z_1 + la)(z_1 - ka)}} \tag{44}$$

and

$$\Psi_2(z_2) = -\frac{pa_1\mu_1}{2\Delta} \left(\frac{z_2^2 - \alpha z_2 - \beta^2}{\sqrt{(z_2^2 - a^2)(z_2 + la)(z_2 - ka)}} - 1 \right)$$
$$- \frac{a_1\mu_1}{2\Delta} \frac{P(z_2)}{\sqrt{(z_2^2 - a^2)(z_2 + la)(z_2 - ka)}}. \tag{45}$$

In this relations

$$\alpha = \frac{k - l}{2}a, \quad \beta = \frac{(1 - l)^2 + 4(1 + l)(1 + k) + (1 - k)^2}{8}a^2$$

From (44) and (45) we can conclude that the boundary condition (16) is satisfied if and only $P(t) = \overline{P(t)}$ for $t \in (-a, -la) \cup (ka, a)$. This requirement shows that C_1 and C_2 from (40) must be real numbers.

According to (4) we have

$$\Phi_j(z_j) = \int \Psi_j(z_j)dz_j, \quad j = 1, 2. \tag{46}$$

To assure the uniformity of the potentials let us denote by (U, V) one of the two cracks. Also let Λ be a simple curve around the crack (U, V) in the complex plane $z = x_1 + ix_2$, as in Fig. 2. Let Λ_j be the corresponding simple curve around the crack (U, V) in the complex plane $z_j = x_1 + \mu_j x_2$, $j = 1, 2$, as in Fig. 2.

According to the relations (1) and (46) the uniformity of u_1 is assured if the potentials $\Psi_j(z_j)$ satisfy the restriction:

$$\oint_{\Lambda_1} \left\{ b_1 \Psi_1(z_1)dz_1 + \overline{b_1 \Psi_1(z_1)dz_1} \right\}$$
$$+ \oint_{\Lambda_2} \left\{ b_2 \Psi_2(z_2)dz_2 + \overline{b_2 \Psi_2(z_2)dz_2} \right\} = 0. \tag{47}$$

Since $\Psi_j(z_j)$ are analytic functions in the complex planes z_j with the cut $\mathcal{L}$, the integrals involved in (48) remain unchanged if Λ is changed. Taking into account this fact and squeezing the curve Λ around the crack, we can conclude that the uniformity of u_1 will be guaranteed if the following condition is satisfied

$$Re \left\{ \int_U^V \left(b_1 \Psi_1^+(t) + b_2 \Psi_2^+(t) \right) dt + \int_V^U \left(b_1 \Psi_1^-(t) + b_2 \Psi_2^-(t) \right) dt \right\} = 0 \tag{48}$$

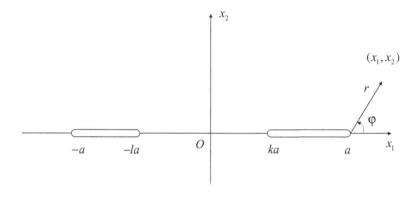

Figure 2.   Asymptotic values near the crack tip a

To obtain the limiting values involved in (48) we use the relations (44), (45) and (37), (38). First we observe that the condition is homogeneous and the replacement of the crack (ka, a) with the crack $(-a, -la)$ implies only the replacement of i by $-i$ in the limiting values. Hence we can consider only the crack (ka, a). In this way from (37) and (41), (42) we get

$$
\begin{aligned}
\Psi_1^+(t) &= \frac{pa_2\mu_2}{2\Delta}\left(\frac{Q(t)}{iA(t)} - 1\right) + \frac{a_2\mu_2}{2\Delta}\frac{P(t)}{iA(t)}, \\
\Psi_2^+(t) &= -\frac{pa_1\mu_1}{2\Delta}\left(\frac{Q(t)}{iA(t)} - 1\right) - \frac{a_1\mu_1}{2\Delta}\frac{P(t)}{iA(t)},
\end{aligned}
\tag{49}
$$

where

$$
Q(t) = t^2 - \alpha t - \beta^2.
\tag{50}
$$

The limiting values $\Psi_1^-(t)$ and $\Psi_2^-(t)$ can be obtained from (49), replacing i by $-i$. Using (49) and returning to the crack (U, V) we get

$$
\begin{aligned}
&\int_U^V \left(b_1\Psi_1^+(t) + b_2\Psi_2^+(t)\right) dt \\
&= \frac{\Gamma_0}{2}\left\{p\int_U^V \left(\frac{Q(t)}{iA(t)} - 1\right) dt + \int_U^V \frac{P(t)}{iA(t)} dt\right\},
\end{aligned}
\tag{51}
$$

where

$$
\Gamma_0 = \frac{b_1 a_2\mu_2 - b_2 a_1\mu_1}{\Delta}.
\tag{52}
$$

As has been shown by Guz (see [5]), and independently by Soós (see [18]), Γ_0 is a real number if the initial deformed equilibrium configuration is locally stable. Taking into account this fact, from (1) and the observation made concerning the limiting values $\Psi_1^-(t)$, $\Psi_2^-(t)$, we can conclude that the uniformity condition (48) is satisfied for both cracks.

We now consider the incremental displacement field u_2. According to the representation formula $(1)_2$ the uniformity if u_2 will be assured if the condition:

$$
Re\left\{\int_U^V (c_1\Psi_1^+(t) + c_2\Psi_2^+(t))dt + \int_V^U (c_1\Psi_1^-(t) + c_2\Psi_2^-(t))dt\right\} = 0
\tag{53}
$$

is fulfilled. Using the same procedure as before we get

$$
\int_U^V \left(c_1\Psi_1^+(t) + c_2\Psi_2^+(t)\right) dt = -\frac{\Gamma i}{2}\int_U^V \frac{pQ(t) + P(t)}{A(t)} dt + i\frac{p\Gamma}{2}(V - U),
\tag{54}
$$

where

$$
\Gamma = \frac{c_1 a_2\mu_2 - c_2 a_1\mu_1}{\Delta} i.
\tag{55}
$$

Guz [5], and independently Soós [18], have shown that Γ is a real number if the initial deformed equilibrium configuration is locally stable. Using this fact, the relation (54) and the observation concerning the limiting values $\Psi_1^-(t)$, $\Psi_2^-(t)$, we can see that the uniformity condition (53) will be satisfied if the condition

$$\int_U^V \frac{pQ(t) + P(t)}{A(t)} dt = 0 \tag{56}$$

is fulfilled. Taking $U = ka$, $V = a$ and $U = -a$, $V = -la$, in (56), we are led to the following two conditions of uniformity:

$$\int_{ka}^a \frac{P(t)dt}{A(t)} = -p \int_{ka}^a \frac{Q(t)dt}{A(t)}, \qquad \int_{-a}^{-la} \frac{P(t)dt}{A(t)} = -p \int_{-a}^{-la} \frac{Q(t)dt}{A(t)}. \tag{57}$$

These above relations will be used to determine the real coefficient C_1 and C_2, appearing in the expression (40) of the polynomial $P(z)$.

To express the system (57) in a simpler form we introduce the following integrals:

$$I_n = \int_{ka}^a \frac{t^n dt}{A(t)}, \qquad J_n = \int_{la}^a \frac{t^n dt}{B(t)}, \qquad n = 0, 1, 2 \tag{58}$$

with

$$A(t) = \sqrt{(a^2 - t^2)(t + la)(t - ka)}, \quad B(t) = \sqrt{(a^2 - t^2)(t - la)(t + ka)}. \tag{59}$$

Now, using the expression (46) for the polynomial $Q(t)$, we can express (57) in the following equivalent form:

$$\begin{aligned} I_1 C_1 + I_0 C_2 &= -p(I_2 - \alpha I_1 - \beta^2 I_0), \\ -J_1 C_1 + J_0 C_2 &= -p(J_2 + \alpha J_1 - \beta^2 J_0). \end{aligned} \tag{60}$$

Now it is easy to see that:

$$C_1 = -p(M - \alpha), \qquad C_2 = -p(N - \beta^2), \tag{61}$$

where

$$M = M(k, l) = \frac{R_2 S_0 - S_2 R_0}{R_0 S_1 + S_0 R_1}, \qquad N = N(k, l) = \frac{R_2 S_1 + S_2 R_1}{R_0 S_1 + S_0 R_1}, \tag{62}$$

and

$$R_n = \int_k^1 \frac{t^n dt}{R(t)}, \qquad S_n = \int_l^1 \frac{t^n dt}{S(t)}, \qquad n = 1, 2, 3 \tag{63}$$

with

$$R(t) = \sqrt{(1 - t^2)(t + l)(t - k)}, \qquad S(t) = \sqrt{(1 - t^2)(t - l)(t + k)}. \tag{64}$$

Using (61) in (44) and (45) we are led to the relations:

$$\Psi_1(z_1) = \frac{pa_2\mu_2}{2\Delta}\left(\frac{z_1^2 - Maz_1 - a^2N}{\sqrt{(z_2^2 - a^2)(z_1 + la)(z_1 - ka)}} - 1\right),$$

$$\Psi_2(z_2) = -\frac{pa_1\mu_1}{2\Delta}\left(\frac{z_2^2 - Maz_2 - a^2N}{\sqrt{(z_2^2 - a^2)(z_2 + la)(z_1 - ka)}} - 1\right).$$

(65)

1.2   Asymptotic expressions. Griffith-Irwin's method

The complex potentials determined in the preceding section will be used to evaluate the energy release rate corresponding to a crack tip. To obtain this quantity we must know the singular parts or the asymptotic values of the complex potentials in a small neighbourhood of the crack tip.

We can get the asymptotic values of the potentials near the crack tip a of the crack (ka, a) by taking (see Fig. 2)

$$x_1 = a + r\cos\varphi, \qquad x_2 = r\sin\varphi, \tag{66}$$

where the polar coordinates r and φ denote, respectively, the radial distance from the crack tip and the angle between the radial line and the line ahead of the crack.

Using (5) and (66) we get

$$z_j - a = r(\cos\varphi + \mu_j\sin\varphi), \quad j = 1, 2. \tag{67}$$

For simplicity we introduce the functions

$$\chi_j(\varphi) = \sqrt{\cos\varphi + \mu_j\sin\varphi}, \quad j = 1, 2 \tag{68}$$

and from (67) we obtain

$$z_j - a = r\chi_j(\varphi), \quad j = 1, 2. \tag{69}$$

We note that in a small neighbourhood of the tip a

$$z_j \approx a, \quad j = 1, 2. \tag{70}$$

Using this observation and the relation (69) near the tip we get the following approximate equation

$$\sqrt{(z_j^2 - a^2)(z_j + la)(z_j - ka)} \approx a\sqrt{a}\sqrt{(1+l)(1-k)}\sqrt{2r\chi_j(\varphi)}, \tag{71}$$

$$j = 1, 2.$$

Now, from (65) we obtain the asymptotic values or singular parts of the complex potential in a small neighbourhood of the tip a:

$$\Psi_1(z_1) \approx \frac{p\sqrt{a}a_2\mu_2m_1}{2\Delta\sqrt{2r}}\frac{1}{\chi_1(\varphi)}, \qquad \Psi_2(z_2) \approx -\frac{p\sqrt{a}a_1\mu_1m_1}{2\Delta\sqrt{2r}}\frac{1}{\chi_2(\varphi)}, \tag{72}$$

where

$$m_1 = m_1(k, l) = \frac{1 - M - N}{\sqrt{(1 + l)(1 - k)}}. \tag{73}$$

To obtain the asymptotic values of the complex potentials $\Phi_j(z_j)$ we use the relations (46). Since the line integrals are path independent we can take

$$dz_j = dr\chi_j(\varphi), \quad j = 1, 2. \tag{74}$$

In this way (72) gives

$$\Phi_1(z_1) \approx \sqrt{2r}\frac{p\sqrt{a}a_2\mu_2 m_1\chi_1(\varphi)}{2\Delta},$$
$$\Phi_2(z_2) \approx -\sqrt{2r}\frac{p\sqrt{a}a_1\mu_1 m_1\chi_2(\varphi)}{2\Delta}. \tag{75}$$

To obtain the critical incremental stress producing the propagation of the tip a we must know the asymptotic values of the normal incremental displacement u_2 and of the normal incremental stress θ_{22} near the tip a. These asymptotic values can be obtained using Guz's representation formulae $(1)_2$ and $(3)_2$, and the relations (72), (75). Elementary calculus lead to the following result

$$u_2(r, \varphi) \approx p\sqrt{a}m_1\sqrt{2r}Re\left\{\frac{1}{\Delta}\left(c_1 a_2\mu_2\chi_1(\varphi) - c_2 a_1\mu_1\chi_2(\varphi)\right)\right\}, \tag{76}$$

$$\theta_{22}(r, \varphi) \approx \frac{p\sqrt{a}m_1}{\sqrt{2r}}Re\left\{\frac{1}{\Delta}\left(\frac{a_2\mu_2}{\chi_1(\varphi)} - \frac{a_1\mu_1}{\chi_2(\varphi)}\right)\right\}. \tag{77}$$

We now consider the tip ka of the crack (ka, a). The same procedure as before leads to the following asymptotic values

$$u_2(r, \varphi) \approx p\sqrt{a}m_2 2\sqrt{r}Re\left\{\frac{1}{i\Delta}\left(c_1 a_2\mu_2\chi_1(\varphi) - c_2 a_1\mu_1\chi_2(\varphi)\right)\right\}, \tag{78}$$

$$\theta_{22}(r, \varphi) \approx \frac{p\sqrt{a}m_2}{\sqrt{r}}Re\left\{\frac{1}{i\Delta}\left(\frac{a_2\mu_2}{\chi_1(\varphi)} - \frac{a_1\mu_1}{\chi_2(\varphi)}\right)\right\}, \tag{79}$$

where

$$m_2 = m_2(k, l) = \frac{k^2 - kM - N}{\sqrt{(1 - k^2)(k + l)}}. \tag{80}$$

In this relations r is the radial distance from the tip ka of the crack (ka, a).
We now determine the asymptotic values near the tip $-la$ of the crack $(-a, -la)$. The same method gives us the following result:

$$u_2(r, \varphi) \approx p\sqrt{a}m_3 2\sqrt{r}Re\left\{\frac{1}{\Delta}\left(c_1 a_2\mu_2\chi_1(\varphi) - c_2 a_1\mu_1\chi_2(\varphi)\right)\right\}, \tag{81}$$

$$\theta_{22}(r,\varphi) \approx \frac{p\sqrt{a}m_3}{\sqrt{r}} Re\left\{\frac{1}{\Delta}\left(\frac{a_2\mu_2}{\chi_1(\varphi)} - \frac{a_1\mu_1}{\chi_2(\varphi)}\right)\right\}, \qquad (82)$$

where

$$m_3 = m_3(k,l) = \frac{l^2 + lM - N}{\sqrt{(1-l^2)(k+l)}}. \qquad (83)$$

To get the asymptotic values near the tip $-a$ of the crack $(a, -la)$ we use the same procedure and in this way we obtain

$$u_2(r,\varphi) \approx p\sqrt{a}m_4\sqrt{2r} Re\left\{\frac{1}{i\Delta}(c_1a_2\mu_2\chi_1(\varphi) - c_2a_1\mu_1\chi_2(\varphi))\right\}, \qquad (84)$$

$$\theta_{22}(r,\varphi) \approx \frac{p\sqrt{a}m_4}{\sqrt{2r}} Re\left\{\frac{1}{i\Delta}\left(\frac{a_2\mu_2}{\chi_1(\varphi)} - \frac{a_1\mu_1}{\chi_2(\varphi)}\right)\right\}, \qquad (85)$$

where

$$m_4 = m_4(k,l) = \frac{1 + M - N}{\sqrt{(1-l)(1+k)}}. \qquad (86)$$

In the relations (81), (76) r is the radial distance from the tip $-la$ of the crack $(-a, -la)$; in (84), (85), r represents the radial distance from the tip $-a$ of the same crack.

As is known, the energy release rates can be obtained using Irwin's relation (see for instance [14]).

In our calculus we take into account the symmetry involved in the problem. Due to this symmetry, we have

$$u_2(t, 0^+) = -u_2(t, 0^-) \text{ for } t \in (-a, -la) \cup (ka, a). \qquad (87)$$

We denote by $G(a)$ the energy release rate corresponding to the propagation of the tip a of the cracks (ka, a). According to Irwin's formula, $G(a)$ can be obtained using the relation (see also Fig. 3):

$$G(a) = \lim_{\delta \to 0} \frac{1}{\delta} \int_0^\delta \theta_{22}(\delta - t, 0)u_2(t, \pi)dt, \quad \delta > 0. \qquad (88)$$

As this formula shows, the energy release rate depends only on the singular part of the normal incremental stress θ_{22} in a small neighbourhood of the tip a.

Since, according to (68)

$$\chi_j(0) = 1 \quad \text{and} \quad \chi_j(\pi) = i, \quad j = 1, 2, \qquad (89)$$

from (74) and (75) we obtain

$$u_2(t, \pi) = p\sqrt{a}m_1\Gamma\sqrt{2t}, \qquad \theta_{22}(\delta - t, 0) = \frac{p\sqrt{a}m_1}{\sqrt{2(t-\delta)}}, \qquad (90)$$

where the real number Γ is defined by the relation (62). Substituting (90) in (88) we get

$$G(a) = ap^2m_1^2\Gamma \lim_{\delta \to 0} \frac{1}{\delta} \int_0^\delta \sqrt{\frac{t}{\delta - t}}dt. \tag{91}$$

Now we use the relation

$$\int_0^\delta \sqrt{\frac{t}{\delta - t}}dt = \frac{\pi\delta}{2}. \tag{92}$$

In this way for $G(a)$ we obtain the following value:

$$G(a) = \frac{1}{2}\pi ap^2m_1^2\Gamma. \tag{93}$$

We use again Irwin's relation to obtain the energy release rate $G(ka)$ corresponding to the propagation of the tip ka of the crack (ka, a). In this way we get

$$G(ka) = \lim_{\delta \to 0} \frac{1}{\delta} \int_0^\delta \theta_{22}(\delta - t, \pi)u_2(t, 0)dt. \tag{94}$$

Equations (76) and (77) give

$$u_2(t, 0) = -2p\sqrt{a}m_2\Gamma\sqrt{t}, \theta_{22}(\delta - t, \pi) = -\frac{p\sqrt{a}m_2}{\sqrt{\delta - t}}. \tag{95}$$

Substituting (95) in (94) we obtain

$$G(ka) = \pi ap^2m_2^2\Gamma. \tag{96}$$

We denote by $G(-la)$ and $G(-a)$ the energy release rates corresponding to the propagation of the crack tips $-la$, $-a$ respectively. Using the involved asymptotic values and the same procedure as before, after elementary calculus we get

$$G(-la) = \pi ap^2m_3\Gamma \quad \text{and} \quad G(-a) = \frac{1}{2}\pi ap^2m_4\Gamma. \tag{97}$$

In order to obtain the critical values of the applied incremental stress for which one of the tips starts to propagate, we shall use Griffith's energy criterion. We denote by $\gamma > 0$ the surface tension of the composite, and we assume that γ is a material constant, having the same value for all tips, and being independent on the initial applied stress.

According to Griffith's criterion (see [14]) the tip a starts to propagate if the condition

$$G(a) = 2\gamma \tag{98}$$

is fulfilled.

Denoting by $p(a)$ the critical value of the applied incremental stress for which the tip a starts to propagate, from (93) and (98) we obtain

$$p(a) = \frac{2}{\sqrt{\Gamma}} \sqrt{\frac{\gamma}{\pi a}} \frac{\sqrt{(1+l)(1-k)}}{1-M-N}. \tag{99}$$

In similar manner we design by $p(ka), p(-la)$ and $p(-a)$ the critical values of the applied incremental stress for which the tips $ka, -la$ and $-a$ start to propagate. Using the obtained results and the same criterion as before we get

$$p(ka) = \frac{\sqrt{2}}{\sqrt{\Gamma}} \sqrt{\frac{\gamma}{\pi a}} \frac{\sqrt{(1-k^2)(k+l)}}{-k^2+kM+N}, \tag{100}$$

$$p(-la) = \frac{\sqrt{2}}{\sqrt{\Gamma}} \sqrt{\frac{\gamma}{\pi a}} \frac{\sqrt{(1-l^2)(k+l)}}{-l^2-lM+N}, \tag{101}$$

$$p(-a) = \frac{2}{\sqrt{\Gamma}} \sqrt{\frac{\gamma}{\pi a}} \frac{\sqrt{(1-l)(1+k)}}{1+M+N}. \tag{102}$$

The numerical tests realised by us show that $1-M-N, -k^2+kM+N, -l^2-lM+N$ and $1+M-N$ are positive quantities for any k and l form the interval $(0,1)$. This fact justifies the choice made by us to express the involved critical value by $(99)-(102)$.

In the same time we observe that the above relations are meaningful only if the real number Γ is a positive quantity. The Eq. (62) show that the values of Γ depend on the elastic constants of the material an on the initial stress $\overset{\circ}{\sigma}$. We assume for a moment $\overset{\circ}{\sigma} = 0$, i.e., a stress-free reference configuration. We also assume transverse isotropy in the $x_1 x_2$-plane.

In this case, using the expressions (11) of the instantaneous elasticities, after elementary calculus we get

$$\Gamma = \frac{C_{11} + C_{12}}{2C_{11}(C_{11} - C_{12})} > 0. \tag{103}$$

This inequality holds since the reference configurations of the material was assumed to be locally stable. From continuity consideration we can conclude that Γ will be positive also if $\overset{\circ}{\sigma}$ is not zero, its value lying in a small neighbourhood of zero. Consequently the relations (99)–(102) are meaningful, even if the initial applied stress is not zero and fulfils the above mentioned condition. Similar results are valuable also for an orthotropic, fiber reinforced and pre-stressed composite material.

However, even in these specified conditions the following question can arise: Is there a critical value $\overset{\circ}{\sigma}_c$ of the initial applied stress $\overset{\circ}{\sigma}$ such that when $\overset{\circ}{\sigma}$ converges to $\overset{\circ}{\sigma}_c$, the coefficient Γ converges to infinity? For a fiber reinforced composite, the answer to this question was given by Guz and it is yes (see [6], Chap. 2). Taking into account the relations (15), for a fiber reinforced compos- ite, Guz was able to show that the critical value $\overset{\circ}{\sigma}_c$ is given by the following relation:

$$\overset{\circ}{\sigma}_c \approx -G_{12}\left\{1 - \frac{G_{12}^2}{E_1 E_2}(1 - \nu_{13}\nu_{31})(1 - \nu_{23}\nu_{31})\right\} < 0. \qquad (104)$$

Since $E_2 \ll E_1$ and $G_{12} \ll E_1$, the critical compression stress $\overset{\circ}{\sigma}_c$ pro- duces only infinitesimal strains in the pre-stressed material. Hence the value of the coefficient Γ can become very large in the domain of infinitesimal strains. In such a situation, the incremental critical stresses $p(a)$, $p(ka)$, $p(-la)$ and $p(-a)$ become very small, and resonance can occur. To avoid such danger- ous situations, leading to the total rupture of the composite, the initial applied compression stress must be drastically limited.

Returning to (99)–(102), we note that values of the critical incremental stresses are given as functions of a and of the dimensionless parameters k and l. This variant is useful for a quantitative analysis of cracks interactions. However, the primary geometrical characteristics of the problem are the cracks lengths and their distance. Thus sometimes it is useful to express the results, in terms of these quantities. To do this we denote the length of the cracks by L_l and L_r and the distance between them by D, as in Fig. 3.

From Fig. 3 it is easy to see that

$$2a = L_l + L_r + D, \qquad a(1 - l) = L_l, \qquad a(1 - k) = L_r \qquad (105)$$

and

$$a = \frac{1}{2}(L_l + L_r + D), \qquad k = \frac{L_r - L_l + D}{L_r + L_l + D}, \qquad l = \frac{L_l - L_r + D}{L_r + L_l + D}. \qquad (106)$$

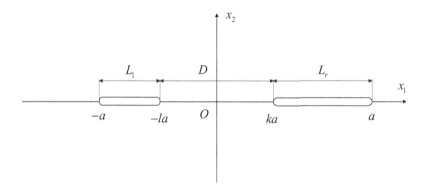

Figure 3. Geometrical characteristics

We shall use these relations in the next section, devoted to the case in which the two cracks have the same length.

1.3   Crack interaction

To test our results, we consider the case of two cracks having the same length L. In this situation the value of the critical stresses are already known for an isotropic material without initial stress (see [17]).

According to the assumption made

$$L_l = L_r = L \tag{107}$$

and from (106) we get

$$a = \frac{1}{2}(2L + D), \qquad k = l = \frac{D}{2L + D}. \tag{108}$$

Since $k = l$, equations (63) and (64) yield

$$R_n = S_n = \int_k^1 \frac{t^n dt}{\sqrt{(1 - t^2)(t^2 - k^2)}}, \qquad n = 0, 1, 2. \tag{109}$$

In this way from (62) we obtain

$$M = M(k) = 0, \qquad N = N(k) = \frac{R_2(k)}{R_0(k)}. \tag{110}$$

Moreover, elementary calculus gives

$$R_0(k) = F(k_1) = \int_0^{\pi/2} \frac{d\varphi}{\sqrt{1 - k_1^2 \sin^2 \varphi}} \geq 1 \tag{111}$$

and

$$R_2(k) = E(k_1) = \int_0^{\pi/2} \sqrt{1 - k_1^2 \sin^2 \varphi} \leq 1. \tag{112}$$

Here $F(k_1)$ and $E(k_1)$ are the complete elliptical integrals of the first and second kind, respectively, (see [17]), of modulus

$$k_1 = \sqrt{1 - k^2}. \tag{113}$$

As the second equation $(110)_2$, shows, N is expressed by the ratio of these elliptical integrals, i.e.,

$$N = N(k) = \frac{E(k_1)}{F(k_1)} \leq 1. \tag{114}$$

Now it is easy to see, using (99)–(102) and the relations (110), that the values of the critical incremental stresses are given by the following formulae

$$p(a) = p(-a) = \frac{2}{\sqrt{\Gamma}} \sqrt{\frac{\gamma}{\pi a}} \frac{\sqrt{1 - k^2}}{1 - N(k)}, \tag{115}$$

$$p(ka) = p(-ka) = \frac{2}{\sqrt{\Gamma}} \sqrt{\frac{\gamma}{\pi a}} \frac{\sqrt{k(1 - k^2)}}{N(k) - k^2}. \tag{116}$$

For an isotropic material and without initial stresses, these above formulae were obtained by Willmore [20], and independently by Tranter [19].

Our formulae generalise these classical results for a pre-stressed orthotropic material.

We assume now an isotropic material, without initial stress considering the problem first analysed by Willmore. Using (11), (12) and (55), after long, but elementary calculus we get the following value for the coefficient Γ:

$$\Gamma = \frac{1 - \nu}{\mu} > 0. \tag{117}$$

In this relation μ is the shear modulus and ν is Poisson's ratio.

Introducing (113) in (111) and (112), we obtain Willmore's classical result (see [17])

$$p(a) = 2\sqrt{\frac{\gamma\mu}{\pi a(1 - \nu)}} \frac{\sqrt{1 - k^2}}{1 - N(k)}, \tag{118}$$

$$p(ka) = 2\sqrt{\frac{\gamma\mu}{\pi a(1 - \nu)}} \frac{\sqrt{k(1 - k^2)}}{N(k) - k^2}. \tag{119}$$

The last conclusion can increase our trust in the correctness of our results in the more complex case of an orthotropic pre-stressed material.

Obviously Willmore's formulae itself can be tested. To do this we assume first that the distance D converges to zero, that is the parameter k tends to zero. In this case k_1 tends do 1, and $N(k)$ converges to zero. Hence from (114) we obtain

$$p(a) = 2\sqrt{\frac{\gamma\mu}{\pi a(1 - \nu)}} \quad \text{if} \quad D \to 0. \tag{120}$$

This is a well-known result, and gives the critical stress corresponding to a single crack having length $2a$ (see [14]).

Now suppose the length L of the crack fixed, and assume that the distance D becomes greater and greater. To obtain the critical stresses corresponding to the limiting case in which D converges to infinity we express the parameters a and D through L and k. Equation (108) yields

$$a = \frac{L}{1 - k}, \qquad D = \frac{2kL}{1 - k}. \tag{121}$$

Introducing the first relation in (118) and (119) we get the expressions for the critical stresses as function of the length L and the parameter k:

$$
\begin{aligned}
p(a) &= 2\sqrt{1+k}\sqrt{\frac{\gamma\mu}{\pi(1-\nu)L}\frac{1-k}{N_0(k)-k^2}}, \\
p(ka) &= 2\sqrt{k(1+k)}\sqrt{\frac{\gamma\mu}{\pi(1-\nu)L}\frac{1-k}{1-N_0(k)}}.
\end{aligned}
\tag{122}
$$

From the second relation (121) we can see that the distance D converges to infinity if and only the parameter k tends to 1. Hence to obtain the critical stresses in the analysed limiting case we must find the limiting value of $p(a)$ and $p(ka)$ when k converges to 1. This problem can be solved using the power series (see [9]):

$$
\frac{2}{\pi}E(k_1) \approx 1 - \frac{k_1^2}{4} + \cdots, \qquad \frac{2}{\pi}F(k_1) = 1 + \frac{k_1^2}{4} + \cdots \quad \text{for } k_1 \ll 1 \tag{123}
$$

Elementary calculus shows that $p(a)$ and $p(ka)$ have the same limiting value. Denoting this common value by $p(\infty)$ we get

$$
p(\infty) = 2\sqrt{2}\sqrt{\frac{\gamma\mu}{\pi(1-\nu)L}}.
\tag{124}
$$

Again, this is a well-known result, and gives the critical stress corresponding to a single crack having length L. Thus we can see that if the distance D between the cracks becomes greater and greater, their interaction becomes weaker and weaker.

These results are plausible and increase our trust in the correctness of Willmore's formulae.

Taking into account this fact we return now to our orthotropic pre-stressed composite containing two equal cracks. In this situation we shall analyse the interaction of the cracks. To do this we return to (115) and (116), assuming a fixed and supposing that the parameter k takes values in the interval $(0, 1)$. Our aim is to establish the relation existing between the critical incremental stresses $p(a)$ and $p(ka)$. For values of k for which $p(a) > p(ka)$ the tips $\pm ka$ start to propagate first. For values of k for which $p(a) < p(ka)$ the tips $\pm a$ start to propagate first. To establish these above order relations we introduce the dimensionless quantities $q(a)$ and $q(ka)$ defined by the equations

$$
q(a) = \frac{\pi a p^2(a)}{4\gamma\Gamma} = \frac{\sqrt{1-k^2}}{1-N(k)}, \qquad q(ka) = \frac{\pi a p^2(ka)}{4\gamma\Gamma} = \frac{\sqrt{k(1-k^2)}}{N(k)-k^2}.
\tag{125}
$$

According to the assumptions made $q(a)$ and $q(ka)$ depend only on $k \in (0, 1)$. The order relation existing between the critical incremental stressed

$p(a)$ and $p(ka)$ is entirely determined by the ratio

$$r(k) = \frac{q(ka)}{q(a)} = \sqrt{k}\frac{1 - N(k)}{N(k) - k^2}, \quad k \in (0, 1). \tag{126}$$

Taking into account the series (119) we see that

$$\lim_{k \to 1} r(k) = 1. \tag{127}$$

To calculate the limit value of the ratio $r(k)$ when k converges to zero, that is k_1 tends to 1, we use the asymptotic developments (see [9])

$$\begin{aligned} E(k_1) &\approx 1 + \frac{1}{2}\left(\ln\frac{4}{k_1} - \frac{1}{2}\right)k_1^2 + \cdots, \\ F(k_1) &\approx \ln\frac{4}{k_1} - \frac{\ln(4/k_1) - 1}{4}k_1^2 + \cdots, \end{aligned} \tag{128}$$

valid for $k_1 \approx 1$. In this way, elementary calculus gives

$$\lim_{k \to 0} r(k) = 0. \tag{129}$$

The values of the ratio $r(k)$ for values of the parameter k in the interval $(0, 1)$ were determined by numerical methods. In Table 1 we give the obtained results for 10 values of k.

In this way we can conclude that

$$r(k) < 1 \quad \text{and} \quad p(ka) < p(a) \quad \text{for } 0 < k < 1. \tag{130}$$

Thus we can conclude that the interval tips $\pm ka$ always start to propagate first. Moreover, since $D/L = 2k/(1 - k)$ we can see that $p(ka) \ll p(a)$ if $D \ll L$, and $p(ka) \approx p(a)$ if $D \gg L$. Hence the interaction between the cracks is strong if the distance between the cracks is much smaller than their length, and is weak if the distance D is much greater than the common length L of the cracks. These results are plausible, being in accordance with observed facts and with the theoretical results given by Kachanov [10] and Slepian [16].

We stress the fact that this presented behaviour of the cracks does not depend on the initial applied stress.

In the next section we shall analyse the behaviour of two cracks with different lengths.

Table 1. Numerical values of the ratio $r(k)$

k	0.01	0.1	0.2	0.3	0.4	0.5	0.6	0.7	0.8	0.98
$r(k)$	0.4	0.8	0.94	0.97	0.98	0.994	0.997	0.9992	0.9997	0.99999

We now return to the general relation (99)–(102) giving the critical incremental stresses for two cracks having different lengths. We assume a fixed and suppose that k and l have values in the interval $(0, 1)$. To study the interaction of the cracks we introduce the following dimensionless quantities

$$
\begin{aligned}
\Pi_1(k, l) &= \frac{p(a)}{\Pi} = \frac{\sqrt{2(1+l)(1-k)}}{1 - M - N}, \\
\Pi_2(k, l) &= \frac{p(ka)}{\Pi} = \frac{\sqrt{1 - k^2)(k+l)}}{-k^2 + kM + N},
\end{aligned} \tag{131}
$$

$$
\begin{aligned}
\Pi_3(k, l) &= \frac{p(-la)}{\Pi} = \frac{\sqrt{(1 - l^2)(k+l)}}{-l^2 - lM + N}, \\
\Pi_4(k, l) &= \frac{p(-a)}{\Pi} = \frac{\sqrt{2(1-l)(1+k)}}{1 + M - N},
\end{aligned}
$$

with

$$
\Pi = \sqrt{\frac{2\gamma}{\pi a \Gamma}}. \tag{132}
$$

Again, to analyse the interaction of cracks we must know the order relation existing between the four critical incremental stresses for different values of the dimensionless parameters k and l. The necessary order relation will be known if the values of the dimensionless quantities Π_1, Π_2, Π_3 and Π_4 are known. These values can be obtained by numerical methods; in Table 2 we give the results obtained by us for 36 values of the pair (k, l).

To interpret these results we use (105) to express the geometrical characteristics L_l, L_r and D of the cracks as functions of a, k and l. We get (see Fig. 3)

$$
L_l = a(1 - l), \qquad L_r = a(1 - k), \qquad d = a(k + l). \tag{133}
$$

From Table 1 we see that if $k = l$, then $\Pi_1 \approx \Pi_4$, $\Pi_2 \approx \Pi_3$ and $\Pi_2 < \Pi_1$. Hence if the cracks have nearly the same length, the internal tips start to propagate first, as we already seen. Let us take $k = 0.001$ and $l = 0.0002$. In this case $L_l \approx L_r \approx a$, $D = 0.003a$ hence $D \ll L_l$. Table 2 gives $\Pi_2 \approx \Pi_3 \approx 0.013$, $\Pi_1 \approx \Pi_4 = 1.6$, hence $\Pi_2 \ll \Pi_1$.

Consequently the critical incremental stresses for which the inner tips start to propagate are much smaller those which can produce the propagation of the external tips. Thus we see that if the distance between the cracks is much smaller than their length, there is a strong interaction between them and they tend to unify.

Table 2.   Numerical values of the parameters $\Pi_1, \Pi_2, \Pi_3, \Pi_4$

k/l		0.002	0.202	0.402	0.602	0.802	0.902
0.001	Π_1	1.577	1.864	1.932	1.957	1.980	1.983
	Π_2	0.012	1.649	1.881	1.974	2.011	2.017
	Π_3	0.015	1.741	2.256	2.901	4.230	6.048
	Π_4	1.673	2.018	2.371	2.972	4.144	5.828
0.201	Π_1	2.007	2.112	2.168	2.198	2.212	2.214
	Π_2	1.745	2.032	2.163	2.226	2.253	2.258
	Π_3	1.648	2.028	2.433	3.045	4.379	6.270
	Π_4	1.872	2.112	2.461	3.021	4.262	5.988
0.401	Π_1	2.361	2.453	2.506	2.168	2.550	2.553
	Π_2	2.260	2.438	2.534	2.585	2.608	2.612
	Π_3	1.880	2.160	2.529	3.128	4.470	2.180
	Π_4	1.938	2.174	2.514	3.078	4.336	6.087
0.601	Π_1	2.914	3.009	3.067	3.100	3.116	3.118
	Π_2	2.909	3.052	3.136	3.183	3.204	3.208
	Π_3	1.972	2.222	2.580	3.174	4.498	6.421
	Π_4	1.970	2.204	2.544	3.110	4.378	6.142
0.801	Π_1	4.123	4.241	4.315	4.358	4.379	4.382
	Π_2	2.248	4.396	4.486	4.538	4.563	4.568
	Π_3	2.008	2.245	2.603	3.195	4.544	6.480
	Π_4	1.985	2.217	2.558	3.126	4.399	6.172
0.901	Π_1	5.789	5.948	6.048	6.106	6.134	6.139
	Π_2	6.120	6.306	6.421	6.487	6.520	6.525
	Π_3	2.014	2.254	2.606	3.199	4.548	6.485
	Π_4	1.980	2.219	2.560	3.128	4.402	6.177

Now consider $k = 0.901$ and $l = 0.902$. In this case $L_l \approx L_r \approx 0.099a$, $D \approx 1.8a$, hence $L_l \ll D$. From Table 2 we obtain $\Pi_1 \approx \Pi_2 \approx \Pi_3 \approx \Pi_4 \approx 6.3$. Consequently all tips start to propagate simultaneously. Consequently if the distance between the cracks is much greater than their lengths, the interaction between the cracks is weak. These results are in good agreement with those obtained before.

Now consider two examples corresponding to cracks having different lengths.

First we take $k = 0.001$ and $l = 0.902$. In this case $L_l = 0.098a$, $L_r = 0.999a$, $D \approx 0.903a$, hence $L_l < L_r$ and $D \approx L_r$.

Table 2 gives $\Pi_1 = 1.983$, $\Pi_2 = 2.017$, $\Pi_3 = 6.048$ and $\Pi_4 = 5.828$. Thus we can conclude that the tips of the longer crack start to propagate nearly simultaneously, and the interaction between the cracks is relatively weak when the distance between them is relatively large, compared with the length of the shorter crack.

As the last example we consider $k = 0.801$ and $l = 0.402$. Now $L_l = 0.599a$, $L_r = 0.198a$, $D = 1.203a$, hence $L_r < L_l < D$. Table 2 gives $\Pi_1 = 4.315$, $\Pi_2 = 4.486$, $\Pi_3 = 2.603$ and $\Pi_4 = 2.588$. Thus we can see that the tips of the larger crack start to propagate nearly simultaneously. The interaction between the cracks is very weak, since the distance between them is relatively large, compared with their lengths.

1.4   Pre-stressed elastic composite with two unequal cracks. Second fracture mode

We consider a pre-stressed, orthotropic, linear, elastic material representing a fiber-reinforced composite material. The symmetry planes of the material are the coordinates planes. The x_1-axis is parallel the fibers. We assume that the admissible incremental equilibrium states of the body are plane states relative to the $x_1 x_2$ plane. The plane containing the cracks is taken as the $x_1 x_3$ plane; the x_2 axis is perpendicular to the cracks faces. We suppose that on the upper and lower faces of the cracks given tangential stresses act. We denote by $\tau = \tau(t)$, $t \in (-a, -la) \cup (ka, a)$, the given tangential stresses.

The components θ_{21} and θ_{22} have to satisfy the following boundary conditions:

$$\theta_{21}(t, 0^+) = \theta_{21}(t, 0^-) = -\tau(t)$$
$$\theta_{22}(t, 0^+) = \theta_{22}(t, 0^-) = 0 \tag{134}$$

for $t \in (-a, -la) \cup (ka, a)$.

Also we have the following conditions at large distances satisfied by the components of the incremental displacements and nominal stress.

$$\lim_{r \to \infty} \{u_\alpha(x_1, x_2), \theta_{\alpha\beta}\} = 0, \qquad r = \sqrt{x_1^2 + x_2^2}, \quad \alpha, \beta = 1, 2. \tag{135}$$

Consequently, from relations (134), (135) we conclude that Guz's complex potentials must satisfy the condition

$$\lim_{z_j \to \infty} \{\Phi_j(z_j), \Psi_j(z_j)\} = 0, \quad j = 1, 2. \tag{136}$$

From Guz's formulas (1)–(3) and from the boundary conditions (134), it follows that the complex potentials $\Psi_j(z_j)$ must satisfy the following conditions

$$\Psi_1^+ + \Psi_2^+ + \overline{\Psi}_1^- + \overline{\Psi}_2^- = 0$$
$$\Psi_1^- + \Psi_2^- + \overline{\Psi}_1^+ + \overline{\Psi}_2^+ = 0 \tag{137}$$

and

$$a_1\mu_1\Psi_1^+ + a_2\mu_2\Psi_2^+ + \overline{a_1\mu_1}\overline{\Psi}_1^- + \overline{a_2\mu_2}\overline{\Psi}_2^- = -\tau(t)$$
$$a_1\mu_1\Psi_1^- + a_2\mu_2\Psi_2^- + \overline{a_1\mu_1}\overline{\Psi}_1^+ + \overline{a_2\mu_2}\overline{\Psi}_2^+ = \tau(t). \tag{138}$$

Now, proceeding as in Mode I, we get the following expressions for the complex potentials (see [1, 2]):

$$\Psi_1(z_1) = \frac{1}{2\Delta} \frac{X(z_1)}{\pi i} \int_L \frac{\tau(t)dt}{X^+(t)(t - z_1)} + \frac{1}{2\Delta} P(z_1)X(z_1) \tag{139}$$

$$\Psi_2(z_2) = -\frac{1}{2\Delta} \frac{X(z_2)}{\pi i} \int_L \frac{\tau(t)dt}{X^+(t)} - \frac{1}{2\Delta} P(z_2)X(z_2). \tag{140}$$

We assume that the tangential incremental stress acting on the cracks faces has a constant value; i.e.

$$\tau(t) = \tau = const. > 0 \quad \text{for } t \in (-a, -la) \cup (ka, a). \tag{141}$$

Using Eqs. (59) and (52) we are led to the following expressions of the complex potentials:

$$\Psi_1(z_1) = \frac{\tau}{2\Delta} \left(\frac{z_1^2 - Maz_1 - a^2N}{\sqrt{(z_2^2 - a^2)(z_1 + la)(z_1 - ka)}} - 1 \right),$$

$$\Psi_2(z_2) = -\frac{\tau}{2\Delta} \left(\frac{z_2^2 - Maz_2 - a^2N}{\sqrt{(z_2^2 - a^2)(z_2 + la)(z_1 - ka)}} - 1 \right). \tag{142}$$

where M and N are given in the relation (62).

Now using these above relations in (67)–(71), (73), (74), (142) and (4) we obtain the asymptotic values of the complex potentials in a small neighbourhood of the tip a:

$$\Psi_1(z_1) \approx \frac{1}{\sqrt{2r}} \frac{\tau\sqrt{am_1}}{2\Delta\chi_1(\varphi)}, \qquad \Psi_2(z_2) \approx -\frac{1}{\sqrt{2r}} \frac{\tau\sqrt{am_1}}{2\Delta\chi_2(\varphi)}, \tag{143}$$

$$\Phi_1(z_1) \approx \sqrt{2r} \frac{\tau\sqrt{am_1}\chi_1(\varphi)}{2\Delta}, \qquad \Phi_2(z_2) \approx -\sqrt{2r} \frac{\tau\sqrt{am_1}\chi_2(\varphi)}{2\Delta}. \tag{144}$$

To obtain the critical incremental stress producing the propagation of the tip a we must know the asymptotic values of the incremental displacement and of the incremental stress θ_{21} near the tip a. Using (1)–(3), (143)–(144) we are lead after elementary calculus to the following result

$$u_1(r, \varphi) \approx \tau\sqrt{2ram_1} Re\left\{ \frac{b_1\chi_1(\varphi) - b_2\chi_2(\varphi)}{\Delta} \right\}, \tag{145}$$

$$\theta_{21}(r, \varphi) \approx \frac{\tau\sqrt{am_1}}{\sqrt{2r}} Re\left\{ \frac{1}{\Delta} \left(\frac{a_2\mu_2}{\chi_2(\varphi)} - \frac{a_1\mu_1}{\chi_1(\varphi)} \right) \right\}. \tag{146}$$

We denote the energy release rate corresponding to the propagation of the tip a of the crack (ka, a) by $G(a)$. According to Irwin's formula (see for instance G. C. Sih and H. Leibowitz [14], J. R. Rice [13] or J. B. Leblond [11]), $G(a)$ can be obtained by using the relation:

$$G(a) = \lim_{\delta \to 0} \frac{1}{\delta} \int_0^\delta \theta_{21}(\delta - t, 0) u_1(t, \pi) dt, \quad \delta > 0. \tag{147}$$

As this formula shows, the energy release rate depends only on the singular part of the tangential incremental stress θ_{21} in a small neighbourhood of the tip a.

From relations (89), (145) and (146) we obtain

$$u_1(t, \pi) = \tau \sqrt{a} m_1 \Gamma_1 \sqrt{2t}, \tag{148}$$

$$\theta_{21}(\delta - t, 0) = \frac{\tau \sqrt{a} m_1}{\sqrt{2(t - \delta)}}, \tag{149}$$

where Γ_1 is defined by relation:

$$\Gamma_1 = \frac{b_1 - b_2}{\Delta} i. \tag{150}$$

Introducing (148), (149) in (147) and taking account of (92) we obtain the following value for the energy release rate corresponding to the tip a of the crack (ka, a):

$$G(a) = \frac{1}{2} \pi a \tau^2 m_1 \Gamma_1 \tag{151}$$

Let $\tau(a)$ be the critical value of the applied incremental stress, for which the tip a starts to propagate; Eq. (98) gives:

$$\tau(a) = \frac{2}{\sqrt{\Gamma_1}} \sqrt{\frac{\gamma}{\pi a}} \frac{\sqrt{(1 + l)(1 - k)}}{1 - M - N}. \tag{152}$$

Proceeding as for the tip a, after elementary but long calculations, we obtain the critical values of the applied incremental stress, denoted by $\tau(ka), \tau(-la)$ and $\tau(-a)$ for which the tips $ka, -la$ and $-a$ start to propagate

$$\tau(ka) = \frac{\sqrt{2}}{\sqrt{\Gamma_1}} \sqrt{\frac{\gamma}{\pi a}} \frac{\sqrt{(1 - k^2)(k + l)}}{-k^2 + kM + N}, \tag{153}$$

$$\tau(-la) = \frac{\sqrt{2}}{\sqrt{\Gamma_1}} \sqrt{\frac{\gamma}{\pi a}} \frac{\sqrt{(1 - l^2)(k + l)}}{-l^2 + lM + N}, \tag{154}$$

$$\tau(-a) = \frac{2}{\sqrt{\Gamma_1}}\sqrt{\frac{\gamma}{\pi a}}\frac{\sqrt{(1-l)(1+k)}}{1+M+N}. \tag{155}$$

For two equal cracks, the values of the critical incremental stresses are given by the following formulae:

$$\tau(a) = \tau(-a) = \frac{2}{\sqrt{\Gamma_1}}\sqrt{\frac{\gamma}{\pi a}}\frac{\sqrt{1-k^2}}{1-N_0(k)}, \tag{156}$$

$$\tau(ka) = \tau(-ka) = \frac{2}{\sqrt{\Gamma_1}}\sqrt{\frac{\gamma}{\pi a}}\frac{\sqrt{k(1-k^2)}}{N_0(k)-k^2}. \tag{157}$$

Our formulae corresponding to a pre-stressed orthotropic material generalize the classical results for an isotropic material without initial stresses.

Now assume an isotropic material without initial stresses. In this case we get

$$\tau(a) = 2\sqrt{\frac{\gamma\mu}{\pi a(1-\nu)}}\frac{\sqrt{1-k^2}}{1-N_0(k)}, \tag{158}$$

$$\tau(ka) = 2\sqrt{\frac{\gamma\mu}{\pi a(1-\nu)}}\frac{\sqrt{k(1-k^2)}}{N_0(k)-k^2}. \tag{159}$$

Suppose the distance D converges to 0, that is, the parameter k tends to zero. In this case we obtain

$$\tau(a) = 2\sqrt{\frac{\gamma\mu}{\pi a(1-\nu)}}. \tag{160}$$

This is a well know result and give the critical stress corresponding to a single crack of length $2a$.

Assume that the distance D becomes greater and greater and the interaction between cracks becomes weaker and weaker. Than

$$\tau(\infty) = 2\sqrt{2}\sqrt{\frac{\gamma\mu}{\pi(1-\nu)L}}. \tag{161}$$

We now return now to an orthotropic pre-stressed composite containing two equal cracks.

We assume a fixed, and suppose that the parameter k takes values in the interval $(0, 1)$. Our aim is to establish the relation between the critical incremental stress $\tau(a)$ and $\tau(ka)$. For the values of k for which $\tau(a) > \tau(ka)$ the tips $\pm ka$ starts to propagate first. For the values of k for which $\tau(a) < \tau(ka)$ the tips $\pm a$ starts to propagate first. To establish this relation we introduce the dimensionless quantities defined by the equations:

$$T(a) = \frac{\pi a\tau^2(a)}{4\gamma\Gamma_1} = \frac{\sqrt{1-k^2}}{1-N_0(k)}, \tag{162}$$

$$T(ka) = \frac{\pi a \tau^2(ka)}{4\gamma \Gamma_1} = \frac{\sqrt{k(1 - k^2)}}{N_0(k) - k^2}. \tag{163}$$

The relations existing between $T(a)$ and $T(ka)$ is entirely determined by the ratio:

$$r(k) = \frac{T(ka)}{T(a)} = \sqrt{k} \frac{1 - N(k)}{N(k) - k^2}. \tag{164}$$

We observe that the relation (163) is similar to (126), i.e. all the numerical results from the Table 2 and their interpretation apply to this case.

We obtain

$$r(k) < 1 \quad \text{and} \quad \tau(ka) < \tau(a) \quad \text{for } 0 < k < 1. \tag{165}$$

Hence the interaction between the cracks is strong if the distance between the cracks is much smaller than their length, and is weak if the distance D is much greater than the common length L of the cracks.

We now return now to the general relations (99)–(102) which give the critical incremental stresses from two cracks having different lengths. We suppose that a is fixed, and k and l have values in the interval $(0, 1)$. To study the interaction of the cracks we introduce the following dimensionless quantities:

$$T_1(k, l) = \frac{\tau(a)}{T} = \frac{\sqrt{2(1 + l)(1 - k)}}{1 - M - N}$$

$$T_2(k, l) = \frac{\tau(ka)}{T} = \frac{\sqrt{(1 - k^2)(k + l)}}{-k^2 + kM + N}$$

$$T_3(k, l) = \frac{\tau(-la)}{T} = \frac{\sqrt{(1 - l^2)(k + l)}}{-l^2 - lM + N} \tag{166}$$

$$T_4(k, l) = \frac{\tau(-a)}{T} = \frac{\sqrt{2(1 - l)(1 + k)}}{1 + M - N}$$

when

$$T = \sqrt{\frac{2\gamma}{\pi a \Gamma_1}}. \tag{167}$$

Again, to analyse the interaction of cracks we must know the order relation existing between the four critical values for different values of the dimensionless parameters k and l.

1.5   Remarks

The main conclusions of this section regarding two cracks in Modes I and II can be summarised as follows.

If the cracks have the same length and this is greater than the distance between the cracks, the inner tips start to propagate first. The cracks tend to unify and the interaction between them is strong.

If the common length of the cracks is much smaller than their distance, all tips start to propagate simultaneously. The interaction between the cracks is weak.

If the cracks have different length, and one of the length is much smaller than the distance between the cracks, the tips of longer crack start to propagate nearly in the same time. The interaction between the cracks is weak again.

There exists strong interaction between two unequal cracks only if the distance between them is much smaller than their lengths. Cracks tend to unify for relatively small values of the critical incremental stresses.

2. Sih's generalized fracture criterion for pre-stressed orthotropic and isotropic materials

Griffith's concept of *energy release rate* assumes the direction of crack propagation to be known *a priori*. Hence, Griffith's theory can treat only problems involving high symmetry. To surpass this difficulty, Sih [15] proposed a new theory of fracture based on the *field strength of local strain energy density*, making a fundamental departure from the classical, concepts. Sih's theory requires no calculation of the energy release rate. Unlike the classical theory of Griffith's energy release rates and Irwin's stress intensity factors, measuring the amplitude of local stresses, the strain energy density factor, representing the fundamental parameters of the new theory, is also *direction sensitive*. Moreover, the *minimum value* of Sih's strain energy density factor represents a new intrinsic *material parameter* whose value at the point of crack instability is independent of crack geometry and loading. In [15], Sih applies his new theory to an isotropic elastic material containing a straight crack inclined at an angle to the loading axis. The theoretical results obtained by Sih agree well with available experimental data.

In what follows we assume the validity of Sih's fracture criterion also for a *pre-stressed orthotropic or isotropic elastic material*. In this case, the usual strain energy density will be replaced by the *incremental* strain energy density.

Later we present an expression for the incremental strain energy density factor corresponding to a straight crack and to the first fracture mode. To do this we use Guz's [6, 7] and Soos results [18], giving the incremental elastic rate corresponding for this problem. We assume that the stress free reference configuration of the material is *locally stable*, and its initial deformation is infinitesimal. We analyse the implications of Sih's generalized fracture criterion for an *orthotropic* and for *transversally isotropic* material, assuming that the body is initially undeformed. We establish the connection between

Griffith's specific energy γ and Sih's new material parameter S_c for an orthotropic material. In all situations, we determine the direction of crack propagation, and compare the results with those existing in the framework of Griffith's and Irwin's classical theories.

2.1   Sih's generalized fracture criterion for pre-stressed orthotropic materials

We consider a pre-stressed, orthotropic, linear elastic material. The coordinate planes are the symmetry planes of the material. We assume that the material is unbounded and contains a right crack of length $2a$, situated in the x_1x_3 plane as in Fig. 4. We suppose that the material is pre-stressed and the initial applied stress $\overset{\circ}{\sigma}$ acts in the direction of the x_1-axis, as in Fig. 4. We assume that the magnitude of the initial applied stress is sufficiently *small*, and produces only *infinitesimal* initial deformations. We suppose that the upper and lower face of the crack are acted on by *symmetrically* distributed normal *incremental* stresses having *constant* value $p > 0$, as in Fig. 4. Finally, we assume that the initial deformed equilibrium configuration of the body is *locally stable*.

Under these conditions the incremental equilibrium state of the material is a *plane strain state*, relative to the x_1x_2 plane. As shown by Guz (see [6, 7]), the incremental elastic state of the body can be expressed in terms of two analytic complex potentials $\Phi_j(z_j)$ defined in two complex planes z_j, $j = 1, 2$. We denote the involved components of the incremental displacement field by u_1, u_2 and the involved components of the incremental nominal stress by $\theta_{11}, \theta_{12}, \theta_{21}, \theta_{22}$.

The expressions for the complex potentials $\Psi_j(z_j)$, $j = 1, 2$, corresponding to the incremental boundary value problem were determined by Guz (see [6, 7])

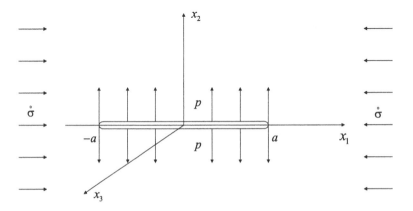

Figure 4.   Pre-stressed crack in Mode I

and later, independently by Soós [18]. We have

$$\Psi_1(z_1) = \frac{a_2 \mu_2 K_I}{2\Delta \sqrt{\pi a}} \left(\frac{z_1}{\sqrt{z_1^2 - a^2}} - 1 \right),$$

$$\Psi_2(z_2) = -\frac{a_1 \mu_1 K_I}{2\Delta \sqrt{\pi a}} \left(\frac{z_2}{\sqrt{z_2^2 - a^2}} - 1 \right)$$

(168)

where Δ is given by (32) and

$$K_I = p\sqrt{\pi a} \tag{169}$$

is the stress intensity factor corresponding to the applied incremental normal stress $p > 0$.

We denote the incremental strain energy density by W; i.e.

$$W = \frac{1}{2} \theta_{kl} u_{l,k}, \quad k, l = 1, 2 \tag{170}$$

where

$$u_{l,k} = \frac{\partial u_l}{\partial x_k}.$$

The expressions for the complex potentials $\Psi_j(z_j)$ show that, near a crack, the incremental strain energy density W has a *singular part* as well as a *regular part*.

To obtain the singular part of W, we denote by r and φ the radial distance from the crack tip and the angle between the radial direction and line ahead the crack.

Starting with the expressions (168) of the potential, and using Guz's representation formulae (1)–(3) we conclude that near the considered crack tip the strain energy density has the following structure:

$$W(r, \varphi) = \frac{S(\varphi)}{r} + a\ regular\ part. \tag{171}$$

Here $S(\varphi)$ is Sih's *incremental strain energy density factor* and is given by the following equation:

$$S(\varphi) = \frac{K_I}{4\pi} s_I(\varphi). \tag{172}$$

Where the function $s_I(\varphi)$ depends only on the elastic constants of the material and on the initial applied stress and has the following expression:

$$
\begin{aligned}
s_I(\varphi) = \\
Re\left[\frac{a_1 a_2 \mu_1 \mu_2}{\Delta}\left(\frac{\mu_1}{\kappa_1(\varphi)} - \frac{\mu_2}{\kappa_2(\varphi)}\right)\right] Re\left[\frac{1}{\Delta}\left(\frac{a_2 \mu_2 b_1}{\kappa_1(\varphi)} - \frac{a_1 \mu_1 b_2}{\kappa_2(\varphi)}\right)\right] \\
- Re\left[\frac{a_1 a_2 \mu_1 \mu_2}{\Delta}\left(\frac{1}{\kappa_1(\varphi)} - \frac{1}{\kappa_2(\varphi)}\right)\right] Re\left[\frac{1}{\Delta}\left(\frac{b_1 a_2}{\kappa_1(\varphi)} - \frac{b_2 a_1}{\kappa_2(\varphi)}\right)\right] \\
- Re\left[\frac{\mu_1 \mu_2}{\Delta}\left(\frac{a_2}{\kappa_1(\varphi)} - \frac{a_1}{\kappa_2(\varphi)}\right)\right] Re\left[\frac{1}{\Delta}\left(\frac{c_1 a_2 \mu_2}{\kappa_1(\varphi)} - \frac{c_2 a_1 \mu_1}{\kappa_2(\varphi)}\right)\right] \\
+ Re\left[\frac{1}{\Delta}\left(\frac{a_2 \mu_2}{\kappa_1(\varphi)} - \frac{a_1 \mu_1}{\kappa_2(\varphi)}\right)\right] Re\left[\frac{\mu_1 \mu_2}{\Delta}\left(\frac{c_1 a_2}{\kappa_1(\varphi)} - \frac{c_2 a_1}{\kappa_2(\varphi)}\right)\right].
\end{aligned}
$$

(173)

In this equation:

$$
\kappa_j(\varphi) = \sqrt{\cos\varphi + \mu_j \sin\varphi}, \quad j = 1, 2.
$$

(174)

Extending the validity of Sih's fracture criterion for *orthotropic* or *isotropic*, *pre-stressed* elastic materials we assume that (see Sih [15])

H1: Crack propagation will start in the radial direction φ_c along which the incremental strain energy density factor as $S(\varphi)$ is a minimum; i.e.

$$
\frac{ds}{d\varphi}(\varphi_c) = 0 \quad \text{and} \quad \frac{d^2 s}{d\varphi^2} > 0.
$$

(175)

H2: The *critical intensity*

$$
S_c = S_{\min} = S(\varphi)
$$

(176)

governs the *onset* of the crack propagation and represents a *material constant*, independent of the crack geometry, loading *and* initial applied stresses.

The assumed independency of S_c from the initial applied stress is similar to a hypothesis made by Guz (see [7, 8]) concerning the independence of Griffith's specific energy γ from the initial applied stresses.

Sih's fracture criterion is based on the *local* density of the incremental strain energy *near* the crack tip and *requires no a priori* assumption concerning the direction in which the energy is released by the separating crack surfaces.

Moreover, according to the relations (169), (172) and (174) the *critical incremental normal stress* p_c for which crack propagation starts at the critical direction φ_c is given by the equation:

$$
ap_c^2 = \frac{4S_c}{s_I(\varphi_c)}.
$$

(177)

In the last relation S_c is Sih's *new* material parameter, which takes the place of Griffith's specific surface energy γ, in the new theory of brittle fracture.

Once S_c, is known, relation (177) can be used to get p_c. The way in which S_c may be expressed through γ for orthotropic materials will be discussed in Sect. 2.3.

2.2 Crack propagation for orthotropic materials

In this section we assume that the orthotropic material is initially unde-formed; i.e.

$$\overset{\circ}{\sigma} = 0 \tag{178}$$

Long but elementary calculations shows that we shall have

$$a_1 = a_2 = 1. \tag{179}$$

Hence, according to Eq. (32) we get

$$\Delta = \mu_2 - \mu_1. \tag{180}$$

The relation (172) giving Sih's strain energy density factor $S(\varphi)$ remains valid, but the function $s(\varphi)$ will have the following simplified form:

$$
\begin{aligned}
s_I(\varphi) &= \\
&Re\left[\frac{\mu_1\mu_2}{\mu_2 - \mu_1}\left(\frac{\mu_1}{\kappa_1(\varphi)} - \frac{\mu_2}{\kappa_2(\varphi)}\right)\right] Re\left[\frac{1}{\mu_2 - \mu_1}\left(\frac{\mu_2 b_1}{\kappa_1(\varphi)} - \frac{\mu_1 b_2}{\kappa_2(\varphi)}\right)\right] \\
&- Re\left[\frac{\mu_1\mu_2}{\mu_2 - \mu_1}\left(\frac{1}{\kappa_1(\varphi)} - \frac{1}{\kappa_2(\varphi)}\right)\right] Re\left[\frac{1}{\mu_2 - \mu_1}\left(\frac{b_1}{\kappa_1(\varphi)} - \frac{b_2}{\kappa_2(\varphi)}\right)\right] \\
&+ Re\left[\frac{1}{\mu_2 - \mu_1}\left(\frac{\mu_2}{\kappa_1(\varphi)} - \frac{\mu_1}{\kappa_2(\varphi)}\right)\right] Re\left[\frac{1}{\mu_2 - \mu_1}\left(\frac{c_1}{\kappa_1(\varphi)} - \frac{c_2}{\kappa_2(\varphi)}\right)\right].
\end{aligned}
\tag{181}
$$

The parameters μ_j, b_j and c_j, $j = 1, 2$ must be evaluated taking $\overset{\circ}{\sigma} = 0$ in the general relations given in the previous section.

In the following we present the results of our numerical analysis concerning the possible values of the critical angle φ_c for two anisotropic materials. For simplicity we have assumed *transversally isotropic materials*, $x_2 x_3$ being the isotropy plane. In this case we have

$$E_2 = E_3, \ \nu_{12} = \nu_{13}, \ \nu_{23} = \nu_{32}, \ G_{12} = G_{13}, \ \nu_{12} = \nu_{31} \tag{182}$$

and

$$\frac{\nu_{12}}{E_1} = \frac{\nu_{21}}{E_2}, \qquad G_{23} = \frac{E_2}{2(1 + \nu_{23})} \tag{183}$$

The *degree of an isotropy* of the materials is characterised by the following ratios:

$$r_1 = \frac{E_2}{E_1}, \qquad r_2 = \frac{\nu_{23}}{\nu_{12}}, \qquad r_3 = \frac{G_{12}}{G_{23}}. \tag{184}$$

If the material is *isotropic* we have

$$r_1 = r_2 = r_3 = 1.$$

To study the dependence of the critical angle φ_c we have considered two materials characterised by the following *fixed* values of the parameters E_2, ν_{23} and G_{23}:

$$E_2 = 5.56\,GPa, \; \nu_{23} = 0.2, \; G_{23} = 2.3\,GPa \tag{i}$$

$$E_2 = 10.3\,GPa, \; \nu_{23} = 0.28, \; G_{23} = 4.02\,GPa, \tag{ii}$$

for

$$E_1 = 75\,GPa, \; \nu_{12} = 0.22, \; G_{12} = 2\,GPa,$$

the set (i) corresponds to a fiber *reinforced aramid/epoxy composite*, and for

$$E_1 = 181\,GPa, \; \nu_{12} = 0.46, \; G_{12} = 2\,GPa$$

the set (ii) corresponds to a fiber reinforced *graphite/epoxy composite*. Both two composites are *strongly* anisotropic.

To study the dependence of the critical angle φ_c, on the degree of the anisotropy of the materials, characterised by the *fixed* sets (i) and (ii) respectively we have determined φ_c for various values of the ratios r_1, r_2 and r_3, taking

$$r_1 \in [10^{-2}, 1], \; r_2 \in [0.6, 1], \; r_3 \in [0.435, 1].$$

Part of the results, corresponding to various combination of the ratios r_2 and r_3, is given in Figs. 5–8 for the first material, characterised by the fixed set (i) and in Figs. 9–12, for the second material, characterised by the fixed set (ii)

From the presented results we can conclude

(a) For $r_1 < 10^{-1}$, *i.e. for strongly anisotropic materials, the critical angle φ_c is not zero*, that is, the direction of crack propagation is not normal

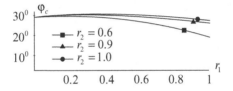

Figure 5. Critical angle φ_c versus r_1 for $r_3 = 0.435$

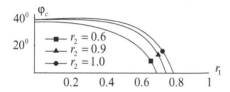

Figure 6.   Critical angle φ_c versus r_1 for $r_3 = 0.87$

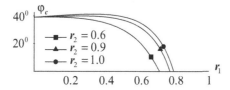

Figure 7.   Critical angle φ_c versus r_1 for $r_3 = 1$

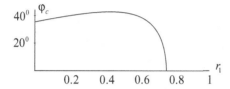

Figure 8.   Critical angle φ_c versus r_1 for $r_2 = r_3 = 1$

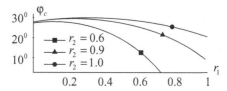

Figure 9.   Critical angle φ_c versus r_1 for $r_3 = 0.5$

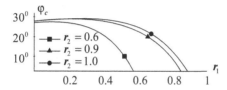

Figure 10.   Critical angle φ_c versus r_1 for $r_3 = 0.75$

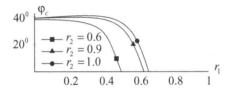

Figure 11.   Critical angle φ_c versus r_1 for $r_3 = 1$

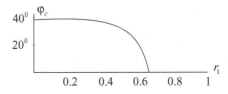

Figure 12.   Critical angle φ_c versus r_1 for $r_2 = r_3 = 1$

to the direction of the applied load, as is usually assumed in fracture mechanics.

(b) The critical angle φ_c is grater r_3 increases and r_2 decreases.

(c) The critical angle φ_c is zero the ratios r_1, r_2, r_3 belong to a small neighbourhood of 1, i.e. when the material is nearly isotropic. For isotropic materials, i.e. for $r_1 = r_2 = r_3 = 1$, we obtain the classical result $\varphi_c = 0$.

2.3   Sih's material parameter S_c versus Griffith's specific surface energy γ

Now assume that for orthotropic material, according to Sih's criterion, the angle φ_c is zero; i.e.

$$\varphi_c = 0. \tag{185}$$

In such a situation, according to the relation (176) we shall have

$$S_c = S_{\min} = S(0). \tag{186}$$

Consequently, if (185) holds, Eq. (181) giving $s_I(\varphi)$ we obtain, (see [3])

$$S_c = \frac{K_I^2}{4\pi} \left\{ -(\operatorname{Re}\mu_1\mu_2)\operatorname{Re}\frac{\mu_2 b_1 - \mu_1 b_2}{\mu_2 - \mu_1} + \operatorname{Re}\mu_1\mu_2\frac{c_1 - c_2}{\mu_2 - \mu_1} \right\}, \tag{187}$$

where

$$K_I^2 = \pi a p_c^2, \tag{188}$$

p_c being the critical normal stress for which crack propagation starts according to Sih's fracture criterion.

Since we have assumed that the stress free reference configuration of the material is locally stable, the parameters μ_1 and μ_2 satisfy the relations.

Moreover, it can be shown (see [6–8] and [18]) that the following equations hold

$$\text{Im}\frac{\mu_2 b_1 - \mu_1 b_2}{\mu_2 - \mu_1} = 0, \qquad \text{Im}\frac{c_2 - c_1}{\mu_2 - \mu_1} = 0 \qquad (189)$$

Accordingly, the expression (187) for S_c can be written in the following equivalent form

$$S_c = \frac{K_I^2}{4\pi}\mu_1\mu_2\alpha, \qquad (190)$$

where

$$\alpha = \frac{c_1 - c_2 + \mu_1 b_2 - \mu_2 b_1}{\mu_2 - \mu_1} \qquad (191)$$

depends *only* on the elastic constants of the considered orthotropic material for which the central angle φ_c is *zero*, according to Sih's fracture criterion.

Let us denote the critical normal stress for which crack propagation starts by P_c; according to the classic Griffith's and Irwin's theory, *a priori* assuming $\varphi_c = 0$. According to Guz's results (see [6–8]) we have

$$K_I^2 = \pi a P_c^2 = \frac{4\beta\gamma}{\mu_1\mu_2}. \qquad (192)$$

where

$$\beta = \frac{C_{12}[C_{11}C_{22} - C_{12}(C_{12} + C_{66}) - C_{22}C_{66}\mu_1\mu_2]\mu_1^2\mu_2^2}{i(\mu_1 + \mu_2)C_{11}(C_{12} + C_{66})}$$
$$+ \frac{C_{11}C_{66}(C_{11} + C_{12}\mu_1\mu_2)}{i(\mu_1 + \mu_2)C_{11}(C_{12} + C_{66})} \qquad (193)$$

is real and depends only on the elastic constants of the material.

Since we have assumed $\varphi_c = 0$, according Sih's fracture criterion, following Sih's (see [15]) for the isotropic case) we suppose that the critical stresses p_c and P_c are equal; i.e.

$$p_c = P_c. \qquad (194)$$

Now, from (190) and (192) we get the relation expressing Sih's new material parameter S_c, through the elastic constants of our *orthotropic* material and Griffith's specific surface energy γ (see [3]):

$$\pi S_c = \alpha\beta\gamma. \qquad (195)$$

This equation is valid only if for the considered material, without initial deformation, we get, according to Sih's fracture criterion for the first mode, $\varphi_c = 0$.

On the contrary, if for this mode Sih's criterion we get $\varphi_c \neq 0$ Eq. (195) is not true. In such a situation Sih's new materials parameters must be determined on experimental way in a first mode fracture test. Once S_c is determined, it can be used to study more complex loading problems.

2.4   Pre-stressed isotropic material

Now consider a *pre-stressed* isotropic material. Let E and ν be Young's modulus and Poisson's ratio, respectively. Now the relation (12) giving the elastic coefficients, take the following simplified forms

$$C_{11} = C_{22} = \frac{1 - \nu}{(1 + \nu)(1 + 2\nu)} E,$$

$$C_{12} = \frac{\nu}{(1 + \nu)(1 - 2\nu)} E, \quad C_{66} = \frac{E}{2(1 + \nu)}. \tag{196}$$

Hence, the instantaneous elasticities, given by Eq. (11), become

$$\omega_{1111} = \frac{1 - \nu}{(1 + \nu)(1 - 2\nu)} E + \overset{\circ}{\sigma}, \quad \omega_{2222} = \frac{1 - \nu}{(1 + \nu)(1 - 2\nu)} E,$$

$$\omega_{1122} = \frac{\nu}{(1 + \nu)(1 - 2\nu)} E, \quad \omega_{1212} = \frac{E}{2(1 + \nu)}, \tag{197}$$

$$\omega_{1221} = \frac{E}{2(1 + \nu)} + \overset{\circ}{\sigma}, \quad \omega_{2112} = \frac{E}{2(1 + \nu)}.$$

To study the dependence of the critical angle φ_c on the initial applied stress $\overset{\circ}{\sigma}$ we must take into account the assumed limits of validity of the model describing the incremental behavior of the material. According to the assumptions made, the initial strain produced by the initial applied stress must be infinitesimal. To assure the fulfilment of this restriction, for the ratio

$$r = \frac{\overset{\circ}{\sigma}}{E} \tag{198}$$

we impose the restriction

$$|r| < 0,02. \tag{199}$$

In what follows we present the results of our numerical analysis for an isotropic material characterised by the parameters

$$E = 5.56 GPa \quad \text{and} \quad \nu = 0,2.$$

Using Sih's fracture criterion for all situations, we have obtained for the critical angle the classically assumed value

$$\varphi_c = 0, \tag{200}$$

as can be seen in Figs. 13–15, for $r \in \{-0,01,0,0,01\}$.

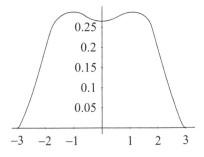

Figure 13. Plot of $s_I(\varphi)$ versus $\varphi \in [-\pi, \pi]$ for $r = -0.01$

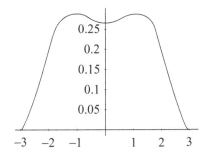

Figure 14. Plot of $s_I(\varphi)$ versus $\varphi \in [-\pi, \pi]$ for $r = 0.01$

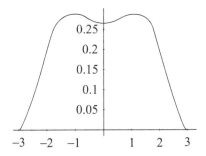

Figure 15. Plot of $s_I(\varphi)$ versus $\varphi \in [-\pi, \pi]$ for $r = 0$

Thus we can conclude that, even if our pre-stressed isotropic material behaves like an anisotropic one, the degree of this apparent anisotropy is weak, and the classical hypothesis used in Griffith's and Irwin's theory is justified, according to Sih's new fracture criterion. Moreover, since Eq. (200) is true, we can use the relation (195) to express Sihs new material parameter S_c through Griffith's specific surface energy γ, and through E and ν.

Long, but elementary calculation based on Eq. (196) show that the relation

$$S_c = \frac{1 - 2\nu}{(1 - \nu)\pi}\gamma \tag{201}$$

holds, as obtained first by [15].

2.5   Sih's energy criterion for the second fracture mode

Consider an unbounded nearly isotropic orthotropic material without initial stresses containing a straight crack having length $2a$. We suppose that the upper and the lower faces of the crack are acted by tangential stresses having a constant and positive value, $\tau > 0$ as in Fig. 16. We assume, as before, that the equilibrium configuration of the material is locally stable.

The elastic state of the body is given by the complex potentials:

$$\begin{aligned}
\Psi_1(z_1) &= \frac{K_{II}a_2\mu_2}{2\Delta\sqrt{\pi a}}\left(\frac{z_1}{\sqrt{z_1^2 - a^2}} - 1\right) \\
\Psi_2(z_2) &= -\frac{K_{II}a_1\mu_1}{2\Delta\sqrt{\pi a}}\left(\frac{z_2}{\sqrt{z_2^2 - a^2}} - 1\right)
\end{aligned} \tag{202}$$

where Δ is given by (32) and

$$K_{II} = \tau\sqrt{\pi a} \tag{203}$$

represents the stress intensity factor in the second fracture mode corresponding to a tangential stress $\tau > 0$.

From the expressions (202) of the complex potentials we see that in a small neighbourhood of the crack tip, the strain–energy–density W has a singular and a regular part.

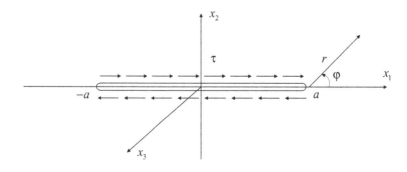

Figure 16.   The crack acted by tangential stresses

From the results obtained by Guz [6–8] and Soos [7] the asymptotic values of the complex potentials (202) are

$$\Psi_1(z_1) = \frac{K_{II}}{2\Delta\sqrt{\pi r}} \frac{1}{\kappa_1(\varphi)}$$

$$\Psi_2(z_2) = -\frac{K_{II}}{2\Delta\sqrt{\pi r}} \frac{1}{\kappa_2(\varphi)}. \tag{204}$$

We obtain the following asymptotic values of the involved strain and stress components involved in the expression of W (see [4]):

$$\theta_{11} = \frac{K_{II}}{\sqrt{2\pi r}} Re\left[\frac{1}{\mu_2 - \mu_1}\left(\frac{b_1}{\kappa_1(\varphi)} - \frac{b_2}{\kappa_2(\varphi)}\right)\right]$$

$$2\theta_{12} = \frac{K_{II}}{\sqrt{2\pi r}} Re\left[\frac{1}{\mu_2 - \mu_1}\left(\frac{b_1\mu_1 + c_1}{\kappa_1(\varphi)} - \frac{b_2\mu_2 + c_2}{\kappa_2(\varphi)}\right)\right] \tag{205}$$

$$\theta_{22} = \frac{K_{II}}{\sqrt{2\pi r}} Re\left[\frac{1}{\mu_2 - \mu_1}\left(\frac{c_1\mu_1}{\kappa_1(\varphi)} - \frac{c_2\mu_2}{\kappa_2(\varphi)}\right)\right]$$

and

$$\sigma_{11} = \frac{K_{II}}{\sqrt{2\pi r}} Re\left[\frac{1}{\mu_2 - \mu_1}\left(\frac{\mu_1^2}{\kappa_1(\varphi)} - \frac{\mu_2^2}{\kappa_2(\varphi)}\right)\right]$$

$$\sigma_{12} = -\frac{K_{II}}{\sqrt{2\pi r}} Re\left[\frac{1}{\mu_2 - \mu_1}\left(\frac{\mu_1}{\kappa_1(\varphi)} - \frac{\mu_2}{\kappa_2(\varphi)}\right)\right] \tag{206}$$

$$\sigma_{22} = \frac{K_{II}}{\sqrt{2\pi r}} Re\left[\frac{1}{\mu_2 - \mu_1}\left(\frac{1}{\kappa_1(\varphi)} - \frac{1}{\kappa_2(\varphi)}\right)\right]$$

Taking into account the expressions (170)–(203), after long but elementary calculus, we obtain the following expression:

$$W(r, \varphi) = \frac{S(\varphi)}{r} + a\ regular\ part, \tag{207}$$

valid in a small neighbourhood of the crack tip.

The intensity of the strain–energy–density field $S(\varphi)$ is given by

$$S(\varphi) = \frac{K_{II}^2}{4\pi} s_{II}(\varphi) \tag{208}$$

where $s_{II}(\varphi)$

$$
\begin{aligned}
s_{II}(\varphi) = {}& Re\left[\frac{1}{\mu_2 - \mu_1}\left(\frac{\mu_1^2}{\kappa_1(\varphi)} - \frac{\mu_2^2}{\kappa_2(\varphi)}\right)\right] \\
& \times Re\left[\frac{1}{\mu_2 - \mu_1}\left(\frac{b_1}{\kappa_1(\varphi)} - \frac{b_2}{\kappa_2(\varphi)}\right)\right] \\
& - Re\left[\frac{1}{\mu_2 - \mu_1}\left(\frac{\mu_1}{\kappa_1(\varphi)} - \frac{\mu_2}{\kappa_2(\varphi)}\right)\right] \\
& \times Re\left[\frac{1}{\mu_2 - \mu_1}\left(\frac{b_1\mu_1 + c_1}{\kappa_1(\varphi)} - \frac{b_2\mu_2 + c_2}{\kappa_2(\varphi)}\right)\right] \\
& + Re\left[\frac{1}{\mu_2 - \mu_1}\left(\frac{1}{\kappa_1(\varphi)} - \frac{1}{\kappa_2(\varphi)}\right)\right] \\
& \times Re\left[\frac{1}{\mu_2 - \mu_1}\left(\frac{c_1\mu_1}{\kappa_1(\varphi)} - \frac{c_2\mu_2}{\kappa_2(\varphi)}\right)\right].
\end{aligned}
\tag{209}
$$

Equations (203) and (208) show that the critical tangential stress τ_c for which crack propagation starts at the critical direction φ_c is given by the following equation:

$$
a\tau_c^2 = \frac{4S_c}{s_{II}(\varphi_c)}.
\tag{210}
$$

Using (195) and (210) we get the following important result:

$$
\pi a\tau_c^2 = \frac{4\alpha\beta\gamma}{s_{II}(\varphi_c)}.
\tag{211}
$$

Griffith-Irwin's theory (see [14]) yields the following relation:

$$
K_{II}^2\mu_1^2\mu_2^2\widehat{m}f^{-1} = 4\gamma
\tag{212}
$$

or in equivalent form:

$$
\pi a\mathcal{T}_c^2 = \frac{4\gamma f}{\mu_1^2\mu_2^2\widehat{m}}.
\tag{213}
$$

Here A and B are given by (7), now from (211) and (213) yield

$$
\frac{\tau_c^2}{\mathcal{T}_c^2} = \alpha\beta\frac{\mu_1^2\mu_2^2\widehat{m}}{s(\varphi_c)f}
\tag{214}
$$

To obtain the relation between τ_c and $\mathcal{T}_c$ we introduce the parameter Q given by

$$
Q = \frac{\tau_c}{\mathcal{T}_c}
\tag{215}
$$

We consider $r_1 \geq 0.8, r_2 = r_3 = 1$, for the material (i) and $r_1 \geq 0.65, r_2 = r_3 = 1$ for (ii)

Table 3. Numerical values of Q versus r_1 for $r_2 = r_3 = 1$

Material	r_1	Q
	0.8	1.01
	0.85	1.02
(i)	0.9	1.04
	0.95	1.05
	1	1.07
	0.65	1.02
	0.7	1.02
(ii)	0.8	1.01
	0.9	1.01
	1	1.00

From these results for two nearly isotropic materials we can conclude that the critical values of the tangential stresses τ_c and $\mathcal{T}_c$ are approximately equal.

In the first fracture mode for the values r_1 taken, the critical angle τ_c is zero. However, as can be seen, τ_c is not zero for the second fracture mode, although the *material* anisotropy is weak. In spite of this fact, Griffith and Irwin's theory and Sih's new fracture criterion lead to *nearly the same values*.

Acknowledgments

The author gratefully acknowledge the support provided to Prof. Tomasz Sadowski – Coordinator of the European Project *Marie Curie Host Fellowships for Transfer of Knowledge (TOK)* – "Modern Composite Materials Applied in Aerospace, Civil and Mechanical Engineering: Theoretical Modelling and Experimental Verification".

References

[1] CRACIUN, E.M. (1999): *Crack propagation conditions for a prestressed fiber – reinforced composite material acted by tangential forces.* Vol. I. Theoretical analysis, Rev. Roum. Sci. Techn.-Mec. Appl., 44, 2, pp. 149–164.

[2] CRACIUN, E.M. (1999): *Crack propagation conditions for a prestressed fiber – reinforced composite material acted by tangential forces.* Vol. II. Numerical analysis, Rev. Roum. Sci. Techn.-Mec. Appl., 44, 4, pp. 435–446.

[3] CRACIUN, E.M., SOÓS, E. (1999): *Sih's fracture criterion for anisotropic and prestressed materials.* Rev. Roum. Sci. Techn.-Mec. Appl., 44, 5, pp. 533–545.

[4] CRACIUN, E.M. (1999): *Sih's energetical criterion and the second fracture mode.* Rev. Roum. Sci. Techn.-Mec. Appl., Tome 45, 6, pp. 663–670.

[5] GUZ, A.N. (1989): *Fracture mechanics of composite materials acted by compression.* Naukova Dumka, Kiev, in Russian.

[6] GUZ A.N. (1983): *Mechanics of brittle fracture of prestressed materials.* Visha Schola, Kiev, in Russian.

[7] GUZ A.N. (1986): *The foundation of the three dimensional theory of the stability of deformable bodies.* Visha Shcola, Kiev, in Russian.

[8] GUZ A.N. (1991): *Brittle fracture of materials with initial stresses.* Naukova Dumka, Kiev, in Russian.

[9] JANKE E., EMDE F., Losch F. (1960): *Tafeln höherer Functionen*, Teubner, Stuttgart.

[10] KACHANOV, L.M. (1974): *Fundamentals of fracture mechanics.* Nauka, Moscow.

[11] LEBLOND, J.B. (1991): *Mecanique de la rupture.* École Polytechnique, Paris.

[12] MUSKHELISHVILI, N.I. (1953): *Some basic problems of the mathematical theory of elasticity.* Noordhoff Ltd., Groningen.

[13] RICE, J.R. (1968): Mathematical theories of brittle fracture. H. Lebowitz (ed.) in *Fracture – An advanced treatise*, Vol. II. Mathematical fundamentals, pp. 192–314, Academic Press, New York.

[14] SIH G., LEIBOWITZ H. (1968): Mathematical theories of brittle fracture. H. Lebowitz (ed.) in *Fracture – An advanced treatise*, Vol. II. Mathematical fundamentals, pp. 68–191, Academic Press, New York.

[15] SIH, G.C. (1973): Mechanics of fracture. G.C. Sih (ed.) in *A special theory of crack propagation*, Vol. I, pp. XXI–XLV, Norhoof Int. Leyden.

[16] SLEPIAN L.I. (1981): *Mechanics of cracks.* Sudostroenie, Leningrad.

[17] SNEDDON, I.N., LOWENGRUB, M. (1969): *Crack problems in the classical theory of elasticity*. Wiley, Inc., New York.

[18] SOÓS, E. (1996): *Resonance and stress concentration in a pre-stressed elastic solid containing a crack. An apparent paradox*. Int. J. Eng. Sci., 34, pp. 363–374.

[19] TRANTER, C.J. (1961): *The opening of a pair of coplanar Griffith's cracks under internal pressure*. Quart. J. Mech. Appl. Math., 13, pp. 269–280.

[20] WILLMORE, T.J. (1969): *The distribution of stress in the neighborhood of a crack*. Quart. J. Mech. Appl. Math., II, pp. 53–60.